An Oriental Infer-System on AL-I & Smart Justice

An Oriental Infer-System on AL-I & Smart Justice

Dr. Shubha Chandra Mishra, Ph.D

GenNext Publication

Published by

GenNext Publication
5, Ansari Road
Daryaganj, New Delhi-110002
Phone: 011-47034999, 9811692060
E-mail: gennextpublication@hotmail.com

Distribution Network
gyanbooks.com
India, USA, Canada, UK, Australia, France

ISBN: 978-93-5663-686-6 (PB)

First Published, 2024

Printed at: Gyan Press, Delhi.

AN ORIENTAL INFER-SYSTEM ON AL-I & SMART JUSTICE

Author: Dr. Shubha Chandra Mishra

Contents

The profile

Name – Dr. Shubha Chandra Mishra
Father's Name –
Vidyavācaspati Late Babu Mishra

Postal Address –
Dr. (prof) Shubha Chandra Mishra
Ward No. 10
P.O. Jhanjharpur Bazar.
Dist. Madhubani (Bihar)-847404
Designation - Principal

M – 9939984914 / 7667921138
Email – drscmishra6@gmail.com

1. Academic qualification – M.A. Acharya. Ph.d.
2. Professional Qualificatioin –
   a. U.G.C Sponsored R.C. Programme — Kurukshetra University Kurukshetra (Harayana) — Qualified- 1991 Qualified- 1991
   b. Do (Auditions performance) — Lucknow university Lucknow (U.P) — do - 1991
   c. do — University of Calicut Kerala) — selected 1992
   d. do — Poona university, Pune (M.S) — Qualified 1995
   e. Professional Development Programme (Officers) — University of Hyderabad, Hyderabad (A.P.) — 2009
3. Seminar/Workshop/Webinar
   a. Bharvi's principle of life (Dais lecture) — -A partial fulfilment of R.C. programme -Poona University Pune (M.S.) — 1995
   b. National ability in Chhatrapati charitam (Dais lecture) — Do — 1995
   c. Workshop on Manuscriptology & Palaeograpy (3 weeks) by Indra Gandhi National Centre for the Arts, New Delhi — In collaboration with Poona University Pune (M.S) — 1995
   d. Research Methodology in Indology — Mithila Institute of P.G. studies & research in Skt. Learning, Kabaraght (Darbhanga) Bihar — 2004

e.	National Seminar sponsored by U. G. C on post-Independence Skt. works on Indian Democracy	B.B.A. Bihar university Muzaffarpur (Bihar)	2007
f.	Learned speech on the great use of social service	Spl. Camp at Dr. J.M.S. College, pastan Navtoli (Madhuban) Bihar	2014
g.	Learned speech on Swami Vivekanand Jayanti & National Youth Day	Kalptaru Cultural Evolutional Trust, Ghaziabad (U.P)	2020
4.	**Project work/Research project**		
a.	Skt. teaching method	A partial fulfilment of the R.C. programe, Lucknow university, Lucknow (U.P)	
b.	An oriental approach to human globe	Indian institute of Advanced Study Rashtrapati Niwas, Shimala (H.P) The project appreciated and fellowship award constrained to limitation-	1997
c.	Minor U.G.C. research project-Supervisor	On Distingguished women in Krishna literature in Mithila Region.- a perception by Dr. Ravindra Nath Jha	2013-14
5.	**Book/Article published/underwork**		
a.	Chhatrapati charitam parichayah	Vishwamaniṣa Bulletin under K.S. D.S.U Darbanga (Bihar)	1981
b.	suddhādwet Ke pravartaka Acharya Ballava ka pustivāda	Mathureshwar jee Felicitation volume Raipur (M.P)	1994
c.	Common culture of Mithila & environment	Mithila institute of Skt. P.G. studies & Research, Darbhanga (Bihar)	2006
d.	Kalsarpayoga Mimānsā	Do (Śastrartha technical Journal)	2008
e.	Research on Sanskrit Literature-Time mandate	"Dainika Jagaran" News Paper (MUZ.) Edition	10-09-2009
f.	Vidyāvācaspati pt. Baba Mishra-A commemoration	Esteemed Journal 'Sāstrārtha' M.I . skt. studies & research Darbhanga(Bihar)	2012
g.	Wine-shun in oriental opinion	Pratibimaba (Hly Journal) M.R.P.L. Manglore (Karnatak)	2015
h.	Awareness compaign for wine-shun	Suggested to Govt. of Bihar	2017
i.	An Oriental Mission for Global Vision	-genNext (Gyan Books) publications New Delhi	2021

6.	**Extra-curricular Activities**		
a.	Chief guest at Education & Cultural Conference on Yuva Avam Loktantra	Under the auspices of Nehru Yuva Kendra, Jhanjharpur (Madhubani)Bihar	1981
b.	Programme officer	Dr. J.M.S College pastan Navtoli (Madhubani)	1994-96
c.	As a Spl. Guest to deliver learned speech on Rastrapitā Gandhi & Village panchayāt	Inaugurated by Hon'ble M.P. (R.S.) M.L. Mandal at P. Raj. Ganauli Andharathadhi Madhubani (Bihar)	2004
d.	Nodal officer (C. unit) (NMEICT)	National Mission for Education through Information & Communication Technology, MHRD, New Delhi.	2010-11
e.	Analyst (Swot & Rusa)-Institutional Development planning & 20-point theme for NEP, MHRD, New Delhi	To the State of Bihar (Educational Deptt.) Patna & MHRD, New Delhi	2014-15
f.	Ex-officio member in University Selection Board. K.S.D.S. University, Darbhanga	Constituted to Dr. R.K.T.J college. Ekahari (Ladania) Madbhubani	2014-16
g.	Ex-officio member (Counselling Committee)	NAAC Peer Team, Banglore for K.S.D.S.U (Darbhanga), Bihar	2016

The Product Feature

In Oriental Holy Writ or Testament, the Karma Philosophy is accountable for all living beings to transmigrate recurrently till salvation. The cessation of life is the consequential event indicating either the achievement of salvation or the succession for the next journey of life. Here the Karma Theory is linked with the logic of cause-cum-effect system being the key-factor in Rebirth. Because at every biomatrix or prenatal chamber, human beings are pledged to lead devout, duteous and solemn life for reverence and benevolence geared to spiritual opulence. In this way they are warranted to work out their pristine knowledge for the progeny. Here the artificial functionality in computation has been the progeny of the themes of this work-product.

But the causticity of time-shift works on the public notion which has been critical and inoperative about oriental learnings since long. We have been apathetic about our archaic knowledge-base due to its inadequacy or incapability to avail excess packages for life-mirth. Since our material life has been persuaded or illustrated to be deluxe to appease carnal appetites to the last breathing.

Although material opulence matters to propel life-cart to a moderate extent, indeed. But when we entertain artificial paraphernalia or belongings, we are made to suffer for multi-agonies as seen or heard in the twilight of life. We do realize the taxing repurcussions of luxuries and when we are left to lament for trials and tribulations towards the ebb of life.

Under the glamour of material opulence in tangible tenure of life, we have no leisure to follow the dedicated focus on cognitive creation of our oriental brains for perennial values.

It is an avowed fact that our life has been bound to suffer transmigration since time immemorial. So we should cultivate our worthwhile and germane life to cast perennial light on the progeny being the receiver or server for untold intelligence. That is why our Alpha Neo-logician Gangeshopadhyaya of Mithila (Bihar, India) proclaimed – The Theory of Syllogism with programming tricks of language about which the present work-product is designed to signify our oriental intelligence. His syllogistic pattern has been seconded by many post neo-logicians of oriental genre. They have deciphered his technical terms with sensible semantics in Sanskrit. Generally his input to output productions are inferred through hierarchical system on the chain of causation enacted by Satkarya-Asatkarya Vāda.

Further the work product has 11 (Nyayic function for electronic design automation) components has contents to infer an accurate conclusion through NN (Navya Nyayic) syllogism. Here the deductive reasoning is applied through 05- correlation as descriptive analysis, inferential analysis, predictive evaluation, prescriptive analysis and deductive determination. Such oriental process in original terms as dealt within the product is justified to workout intended results in the given information. Here the knowledge-base (pakshata) refers to the penta-processor of analyzing and stimulating logical tasks for output transmission. The amount of AL-I (Artificial Intelligence) deals with making predictions to carry out various tasks through Inference-Engine without human intervention physically.

The AL-I Theory is the by-product of Oriental Theory of Neo-logic envisaged in the 13th (14th) century. NN Gangeshopadhyaya formulated AL-L (Artificial Language) for programming communications. His techniques of assembling or arranging error-free and squarely precise programme for task-touch are treated in data-stores and word segments. He has stressed the application of NN syllogism which works on inferential system in the mechanism of simulation through Bio-tools. His own programming Methodology for task fruition is unlike NL (Natural Language).

Although his modus operandi logistics on AL-I is something twisted and terse to understand instantly. But while decoding the deep learning (DL) of his software designing, one finds his automated strategy to be argumentative and without any fallacy of reasoning. So the work-product is conducive to qualify his brain-browsing programme with highly creative intelligence. To be more clear, his inferential data-evaluation undergoes through penta-tools on the concept of invariable concomitance (vyapti) between middle term and major term. His Inference Rules function through premises, syntax analysis and conclusion or Judgement. This school of Neo-logic works on syllogism with argumentative hierarchy as proposition (Sadhya), Reason (Hetu), Illustration (Udaharan), Application (Upanaya) for others'- parārthānumān – the mountain has smoke which is invariably concomitant with fire – to bring the argument nearer to the truth-conclusion (Nigamana). The work product befits us to evaluate dialectical concepts in order to avert fallacy of reasoning in an authentic genre.

As we face with probabilistic chances of dilemma in daily life, We should cultivate such logical formula or equations for valid deduction. Such practice would enhance our critical thinking-skill with personal experience and recollection. It may be helpful to enrich our complex-skill to get at the eventuality validly or evidentially.

Such inferential strategy is useful to involve evidence and reasoning with prior experience to avail unknown (need) facts wholly. Such inferential knowledge through NNM (Navya Nyaya Model) is fault-free and unfallible through observations and logical arguments in inducto-deductive database.

The Functional Note

Since time immemorial, the geological or geographical sustainability in view of habitable survivals or vicissitudes of the subcontinents entails statehood boundaries for socio-economic running of life. So the macro to micro and viceversa landmarks are let navigated for habitable life in exigency. After centuries lapses we may treat such territorial borderlines as own or the other extent causing argumentative disputes on speculations. To me one should not tease over match-mismatch instinct rather esteem the majesty of the sub-continental lands.

Although naturally one knows one's land chronicles better to deal for the progeny. I mean to say that the land of erstwhile Mithila delivered amazing phi-icons in the field of Natural Science, Cognitive Science, Space Science, Natural Philosophy, Meta-Physics, Telemechanics, Linguistic, Literature, Aesthetics, Theology, Artificial terminology and inter-intra debate science. Especially the land of Mithila is acclaimed to be the seat of ancient and modern logicians/philosophers: Like –

Ancient Period

Acharya Akshapad Gautam – Nyayasutra

Acharya Vatsyāyan – Nyayabhasyakār

Acharya Kapil – Sānkhya Darshan (cosmological exponent)

Acharya Kanad – Natural physicist

Acharya Prashastapād – Prashastpad Bhashya exponent on Nyaya Vaisheshik

Acharya Vdyotkar – Nyaya Vartik – Critic

Medieval Period

Acharya Dignag (Buddhist) – Pramān Samuchaya

Acharya Dharma Kirti (Buddhist) – Pramāna Vartika

Acharya Jaimini – Purva Mimansa, Jaiminisutra

Acharya Kumaril Bhatta – Shloka-vārtik, Tantra-Vartik on Shabarbhāshy

Acharya Mandan Mishra – Mimansanukra manika and Vibhrāmviveka

Acharya Prābhākar Mishra - Mimansaka – Shābarbhāshyatika

Acharya Trilochana – Naiyayic (logician)

Acharya Vācaspati Mishra – Nyaya Tatparyatikakar (Rephraser)

Acharya Keshav Mishra – Tarkbhāshā

Acharya Shaliknath Mishra – Naiyayic

Acharya Parthsarthi Mishra – Shāstradeepika

Acharya Padmanābha Mishra – Logician

Acharya Misaru Mishra – Vivadchandra

Acharya Ayachi (Bhavanath) Mishra –Naiyayik Expert

Acharya Shankar Mishra – Kanād Rahasya

Modern Period

Acharya Udayan – Nyaya Kusumānjali,, Nyaya Vartiktāttaparya Tikā Parishuddhi

Ātmatattva Viveka, Kirnavali

Raja Mahesh Thakur – Scholar Prince

Acharya Gangeshopadhyaya – Tattvacintāmani

Acharya Yadunath Mishra – Gudhārtha deepika

Acharya Keshav Mishra –

Ruchipati Upadhyaya –

Pakshadhar Mishra –Tattvachintamani "Aloka"

Acharya Gokulnath Upadhyaya –

"Avayabtattvachintamani"

"Didhitividyot Padvakya Ratnakar

and many more.

Besides other phi-icons who shared textual talents with Mithila erudition chiefly are:

Acharya Vasudeosārvabhoum – Mnemonic Expert Founder-Tattvachintamani "Vyakhyā"

Acharya Raghunath Shiromani – Tattavachintamani "deedhiti"

Acharya Vardhaman Upadhyaya – Tattavachintamani "Prakash"

Acharya Gadādhara Bhattacharya – Vyutpattivadah Anumāngādādhari on Tattvachintamani and deedhiti

Avayava Prakaranam and many more.

Acharya Jagadish Tarkalankar - Anuman Jagadishi on "Deedhiti"

Acharya Mathuranath Tarkvagish – "Vyaptipanchakarahasyam".

Under the impetus of the iconic flow of the land, I, being the chip-chain of phi- scholar lineage, happened to harness the cogency of conscience of my celebrated brain-browsers. That is why I successeded to publish my Maiden cyclopedia, "An Oriental Mission for Global Vision" in 2021. Unlike ultramodern mindset I am eager to focus on feeding oriental intelligence via inferential mechanism geared to tooled intelligence and offended trial for appeasement in life.

In order to decode our oriental cognition in days of Mechatronics, I should opt Indian Syllogistic or Mathematical model for argument procedures which are relevant in today's technology. Because mathematical proofs are logical and systematic argument affecting false or true statement. Because proofs by deduction from a known fact or definition moves in logical steps towards final conclusion. As my concern is to treat the Nyayology which covers the science of reasoning, epistemic relevation, systemic investigation, Psychoanalytic criticism, Metaphysical perception, reductio-argument and analogio-critical analysis resulting valid or crystal cognition or conclusive judgement on trial for human favour. Further Nyayology acts as a pinion gear to synchronize self-mind and mind + sense-organs. Because the intent of human beings is to aspire for material as well as immaterial endowment for Beautitude.

Indian philosophy (Theistic and Atheistic) deals Cosmology, Metaphysics, Epistemology and Ethics. The reason behind the emergence of Indian philosophy is to abstain from human sufferings or miseries or wanderings for want of legitimate knowledge. In order to avail the core-link of legitimate knowledge, one should undergo the gateway of Nyaya philosophy to the extent of Navya-Nyaya (New logic or the science of reasoning) to satisfy the essential needs of the wise public. It may be called cognitive testimony or reasoning-prone learning–the science of logical organon. It enlightens, sharpens and magnifies our Brain. Whereas human beings in general are problematic in validating among miscellaneous views where NN (Navya Nyaya) is instrumental to get at the crystal efficiency of human intelligence.

Towards the end of the 12th century. Acharya Udayan (Udayānachārya) on quashing the impugned arguments of Buddhist philosopher (guānshree) hit his micro-induction on Nyaya – Vaisheshic treatise synchronized "Ātmatattvaviveka". Wherein he seeded the thesis of NN (Navya Nyaya) as quantum mechanics within logical matrix or substratum per se in 09 entities to sustain the existence of primal elements (Earth, Water, Fire, Air, Ether, Tune, Spirit, Space and Mind) and the rest of which are –

(a) Property (guna) – which is subdivided into 24 species
(b) Action-cause resultant into its effect or Natural Law of the Universe.

{Here Action & property are genetic into substance.}

(c) Communion or genus –denoting similarities that cause two or more objects to be classified simultaneously.

(d) Inherence –which implies objects coupled naturally. Such gnoseological axiom is based on the correlation of our sensory channels.

So the Naiyayikas exercised their intelligence to innovate the scientifice credo of logical argumentals and observations through 04 organs of legitimate knowledge viz. Perception, Inference, Analogy and Verbal testimony. And we would be able to deal with puzzling or baffling problems in a systematic sequence. Further NN (Navya Naiyayikas) penned treatises on mathematical reasoning, philosophical syntax, semantics and micro-nature of the phenomenon.

Here I am inclined to explore the Neo-Logic Theory of legitimate knowledge based on infallible cognition under causal condition of inference vouchsafed by Alpha Neo- Logician Gangeshopadhyaya's (Gangesh's) monumental work –Tattvachintamani and its commentators-Gadādhar Bhattacharya, Ragunath Shiromani and the like. He (Gangesh) created a new trend of artificial diction on the clear-cut exposition of thoughts. This Neo-Logic (Navya Nyaya) is the server of computer oriented tool to computational linguistics for programming language. It can work out the issue of anagram or ambivalence on the conflict of human interactions. This unique work of Gangesh is of epoc-making nature in the realm of Neo-Logic in the Medieval to modern period. The work proved to dispel the deep nescience of orthodox thinkers wittily and artfully.

NN is a world's wonder in the realm of wisdom and innovation into wide sphere of artificial intelligence. Oriental philosophers pioneered to acknowledge "Inference" as a specific tool to determine legitimate knowledge. In our daily life, we make decisions about issues through trials on speculations, predictions, implications and hypothesis in a given situations or informatics. Whereas –

(a) Speculation – is to muse, meditate and mull over events in a tentative way.

(b) Prediction – to forecast on existing events out of experience or discuss about future events or something that can be explicitly verified within natural cause of anything.

(c) Implication – to convey the idea indirectly in an inaccurate way.

(d) Hypothesis – to provide tentative declaration or assertion or communication in an approximate way.

Whereas the Inference (Anumān) is a tool to draw logical conclusion or generalization with deductive reasoning, investigation or review or reflection on preceding or previous realization or cognition as conditional in brief.

Further the Theory evolved around implicit analysis based on universal evidence and clues more than just assumption. In a simple language, we use Inference in any moment – if we infer that some events have taken place elsewhere of which we could not see, feel, smell or taste the real causation. Still we make inferences false or true from the same or separate sources. Besides off beaten track, we may come across the utility of inferential mechanism in – Theology viz –

(a) Kātyāyan Smriti – discerning the accurate intention to infer on external manifestation i.e. sweat, piloerection.

(b) Manu Smriti – quoting hunter finds out den or tunnel of an injured deer inferred by drops of blood to discover by the king.

(c) Vyakaran – An application of suitable formulas in word-formation on suggested inference.

(d) Dramaturgy – The 7th out of the 13 elements of the dramatic juncture or catastasis inferred by the Analogy.

(e) Rhetorics – related to cause-effect variable concomitance to beautify their literary composition by Acharya, Mammata, Vishwanath, Jaydeo etc.

(f) Ayurveda (Oriental Life Science) – hypothetical test-trials on inferential evidence to adjudge the potency of production. Further unseen findings are inferred on the succeeding observations.

(g) Taittariya Brahman Granth – Utility of inference, pervasion and conclusion (judgement) are seen to have practised in scattered ways.

(h) Specific difference – is the key-conception of logic wherein structural essence distinguishes the substance from other things from the same genus. Here the fundamental truth of wisdom warrants natural relationship to the world and among others.

(i) Negation – (post added) Here negation as a substance has a sense of positive. Because absence is not a separate means of knowledge. It is also perceived through senses opposed to Mimansak Kumaril Bhatta's views and Vedantins. A man has a perception of an absence of missing somethings. Absence may be of 04 types

   (i) Previous absence - of a new product.

   (ii) Later Absence – of a destroyed object.

   (iii) Total absence – as of colour in the mind.

   (iv) Mutual or Common Absence – as of a pitcher (Ghat) and a cloth (pata) – neither of which is the other.

As my enterprise is to synthesize syllogistic system of thought, belief, emotion, image and locomotion to observe and judge in accuracy through oriento-reliable system helps to analyze, evaluate and inferential model. Such theoretical formulas or devices working on functional automation are integrated to infer valid solutions or results in complex reasoning. Such oriento-reliable system helps to analyze, evaluate and integrate accurate predictions on data-base of input through syllogistic rules or features.

Lastly the Neo-Nyayology is systematized in sequential model tested to deductive accuracy on past experience and recollection reflected in natural mind or inner organs.

Knowledge & Valid Knowledge

Knowledge

Knowledge means information or consciousness or awareness or perception or cognizance or recognition of objects. It consists of all recognitions that have more or less determinative to objective references.

The object of cognizance may be a thing or a quality or an act or emotions – existent or nonexistent as well.

(a) Further knowledge reveals our thoughts, feelings and beliefs to our physical reality. Knowledge is a mere manifestation of objects.

(b) Knowledge may be said to be the illuminating feature of the Self. Without such luminous light we may be deprived of all rational practices and creative activities. It is on the basis of even a bit of behavioural knowledge that we are able to deal with other living objects around the world.

(c) Knowledge is an intellection differs from appreciation or affection or optional preference.

(d) Knowledge is not a raw material or a composite cause of anything nor is the constant cause of precise recognition with variant feature.

(e) Knowledge is perceived or observed by the inner sense called mind. Whereas physical features are observed through their sense organs.

(f) Knowledge relates to all cognitive causes alike–

(i) Sensation

(ii) Perception

(iii) Inference

(iv) Memory

(v) Suspicion

(vi) Dream

(vii) Illusion and the like

In oriental Ideology, knowledge need not be linked as feature with other physical existence –Matter, Mind and Self. Metaphysically it is the vital model of the sub-division of the universe to be realized with self-luminosity of percipience.

Knowledge may be:

(a) Psychological – relating to the interpretative diagnosis of the knowledge.

(b) Logical – critical interpretation of the modus operandi of knowledge.

(c) Philosophical – conclusive trial to validate knowledge to fathom out the reality.

Remarkable to note here that these analyses are tough to be detected from one another in the event of abstract classification.

Valid Knowledge

All cognitions are not valid knowledge. In order to acknowledge the nature and norms of valid knowledge, we, firstly, should study the science of NN (Navya Nyaya) of Indian philosophy. Indian Logic is not based on faiths and superstitions, rather adjudges syllogistic norms on logical analysis and evaluation of events. It is the natural science to investigate the logical methodology to conclude the valid source of knowledge. Further it testifies the validity of knowledge on prior experience of universal pervasion through induction to deduction.

Acharya Akshapada Gautam states – when the true nature of the condition of motive or cause is not decided or adjudged, one should exercise reasonable observation arguing pros and cons of the topic or event discussed deductively or reasonably (ooh) is called Nyaya (Indian Logic). Further it defines the identification of pervasion (Vyapti) by that of the artificial or external property of the middle term to be pervaded on the relation of the inherent property is called Nyaya. It may be fallible on the negation of 03 technical parameters –

(a) Ashrayāsiddhi (unwarranted premises) – negation of determinant fact or characteristic in the knowledge-base or logical subject as heaven lotus-here the negation of adjectival cause of (fragrance) of sky-lotus persists in the knowledge-base (pakshata).

(b) Sadhayāprasiddhi (non-conclusive major term) – the negation of the intended or essential nature in the syntax.

(c) Ashrayālinga – the negation of an adjectival Mark signifying the word which it qualifies as distinctive character.

So the science of critical study perceives general to particular or particular to general object to the extent of intuitive perception through subtle mechanism. Because Indian Logic attempts to argumental running into natural science and does not related to material truth merely.

The Nyaya-Vaisheshika philosophy admits Inference being the source of knowledge. So inferential knowledge is a valid knowledge ushered by invariable concomitant of the probans in the logical base. Because Inference generates Inferential knowledge when the probans is bestowed with pervasion in the logic-subject.

Further the inferential knowledge is gained by the universal relation (Sāmānya laxan) between the probans and the probandum as unconditional. Here the logicians contend that the previous knowledge of the pervasion in all time and space plays a crucial role to ascertain the probandum in the subject. The observation of the invariable concomitance serves a fulcrum to steer up the accredited source of valid knowledge.

In another words the absence of the determination of probandum qualified by the absence of desire to identify the probandum–is termed as inferential knowledge (valid knowledge).

Here it is pertinent to add that logician Lougākshi Bhāskar in his booklet "Tarka Kaumudi" explained lucidly that inferential knowledge as a valid knowledge being the wherewithal of invariable concomitance by casting off the state of occasioning the probans (lingagyān) in the knowledge-base. Further valid knowledge is conceived through rungs of logical parameters with technical upstairs as –

(a) Pakshatā – Where the relative token or identity is being perceived.

(b) Pakshadharmitā – The presence of the relative logo or token in the logical subject.

(c) Ling parāmarsha – The recognition of the existence of the incidental knowledge remotely or distantly on the basis of the perceptual knowledge.

(d) Vyāpti – The knowledge of the inherent relation between the probans and the probandum on the premises of the priori perceptual knowledge navigated by memory.

(e) Anumiti – The concluding recognition of the existing knowledge in the major term.

Now one may happen to come into a ditch of illusionary knowledge reversed of valid knowledge being the bestowal of a substance in its genuity.

Whereas of illusionary or invalid knowledge is the misidentification of a substance dubiously. Here the substance in Viparyaya is mistaken for its having feigned character or property beyond its physical existence. The phantom of awareness of a substance is received against its real character as follows:-

(a) the comprehension of a conch shell is mistaken for a sterling object.

(b) a cord mistaken for a serpent.

(c) a post mistaken for a person.

(d) a mirage mistaken for water.

In all instances the recognition is misviewed against its real status.

So such illusioned or erring recognition is invalid knowledge.

Here such realization is different from doubt. Because it is not only invalid as positively invalid as false conclusion through feature or property suspended on each event.

Mirage

The recognition appears to be that of water instead of sunrays. Though there is no error in the plain comprehension of the object. The object as given in indeterminate perception (Nirvikalpaka gyān) has a series of really present wobbling rays of the sun. Here the given data are mistaken for water in its determinate (Savikalpaka) perception.

In the Nyaya Theory – There is the prima facie contact of sense-organs with some substances. The sense-organ overlooks or reneges its individual or specific character. But the vague traits with something silver or glittering object) remind the memory image of the distinctive character of other thing. There may be infirm or dim contact between sense organ and other thing.

Hence fallacy or delusion lies not in appearance or exposition of involved perception rather in the determination of one's appearance by the other through correlated memory. Since such conviction results in the conclusive outcome of the substance as opposed to its distinctive character. Nyaya Theory calls it Anyathākhyāti or Vipritkhyati (fallacious conception). Such uni-perception is incompatible to the Mimansaka.

Prabhakar's multiplicity of perception and memory with their significance is dimmed by arcana of recollection. Accordingly a fallacious cognition is representative in character with some base. But the fact being misplaced and misrelated – fallacy turns into an unreal comprehension of the real.

Here the error of cognizing a glittering object is an error of perception and not of recollection. Consequently even Anyathākhyāti Siddhānta (fallacious conception theory-philosophy) evades evidence.

Whereas Naiyayikas interpret as a type of sense-object contact caused by presumptive renascence of the past experience of a substance. So the perception of a glittering object (silver) does not appear as an idea or image of the mind. But it is the cause of contiguity presented by the sensing of bright colour. It is the presentation of a silver in a wrong affinitive relation conferring illusive knowledge.

Further we have sources of scientia as illustrated in Oriental Logic:-

(a) Perception (Pratyaksha) – that which is before one's eyes to have immediate cognition within the Globe. Further it is cognitive immediacy linked with head-on perception on recollected sources. It is external through sensory organs.

(b) It may be internal to be experienced within the Self inwardly.

(c) It may be called non-discriminatory perception in which an object is perceived without its codifying feature.

(d) Discriminating in which both codifying features (perceived + recognized) are realized.

Further this perception is of 02 types:-

(a) Usual or customary- when sense organ is linked with the object in an usual way. Again this usual perception maybe of 02 types:-
    (i) External – when 05 sense organs come into contact with external object.
    (ii) Internal – when our mind comes into contact with psychological states and process like – cognitive affection= conation, desire, pain, pleasure, aversion and etc.

(b) Extraordinary or unusual - if the contact of sense-organs occurs in an unusual way having 03 facets:-
    (i) Universal perception
    (ii) Complex perception through contact
    (iii) Yogic property (Yogaj Laxan) – the objects are not directly presented to sense organs. But they enkindle into memory through prior cognition and are perceived through rendition. Such is intuitive and immediate perception of all objects – past, present and future through the power of meditation. It is supra-sensuous and supra-relational.

(02) Inference – (Anumān) – It is the mediate and indirect source of knowledge through penta-tool of syllogism based on valid probans reasonsmoke) and probandum (natural reason) as cognitive validators. It serves as epistemology to find out conclusion or judgement on facts, informations and evidences through logical introspection and investigation. It argues inductive to deductive results drawn from two or more premises through mediate observation and trial of reasoning.

It has 02 stages:-

(a) One pertaining to the preceding (known)

(b) The other pertaining to the succeeding (unknown)

Both have a special kind of correlation (ling-lingibhāva) to infer. The succeeding is bound to come from the preceding as both are taken in invariable concomitance under the cause-logical subject. So the union of the both logical subjects with universal concomitance involves inferential knowledge due to causal condition.

(3) Analogy (Upamāna) – Oriento-logician Udayanācharya in "Nyayakusumānjali" and Vishwanath Panchanan in "Bhāshāparichheda" opine that we gather about valid (new) knowledge through resemblance (prior knowledge). As the preceding knowledge of the feature of domestic cow helps to recognize the wild cow (succeeding knowledge) by virtue of cowhood cognity on recollection. Here we have a glimpse of cow-like animal in the forest to ascertain the succeeding knowledge of the wild cow through resembling operation.

In other way we recognize earthen pot of remote place with prior knowledge of the same sort of pitcher at home by virtue of resembling operation from time immemorial. Whereas the recent norms of relationhood among entities entail further definition.

Here recollection differs from valid and invalid experience to guess extra-error in causality. Again recollection or memory or reminiscence generates quasi knowledge on the plea of prior experience through awakening of accomplishment.

(4) Verbal Testimony (Shabda) – It is the valid source of knowledge defined as a credible or reliable utterance of thought to know its meaning. A word or a recollection of words has potential symptom to convey its meaning based on memory or flashback or recognition or identification. The directly expressed meaning of a word depends on causal character to validate signification of universal nature. A word is defined as utterance of a reasonable for credible person who has the way in which his mind, action and speech are let harmonized to lead as authentic and sincere of the nature of the perceptible objects.

A person might be similar (Sapaksha) in look but dissimilar (Vipaksha) in nature. In dilemma one might gather related terms with specific and semantic feature of the locus.

Thus is the basic intro of Nyayology in brief. Exhaustive interpretations of its technical operation are dealt in relevant chapters ahead.

NN Gangesh and Brain-gem-browser

As my concern is to optimize the mechanism of the Neo-Logic School of Indian Philosophy, I do admire NN (Navya Naiyāyika) Alpha Acharya Gangeshopadhyaya of erstwhile Mithila region where he explored his intuitive talent in "Tattvachintāmani" (Brain-gem-probe) where is his finesse explored adroit and Artificial language to step-up, scan, interpret, integrate and work out technicality in Neo-Logic like real gem as his brain-gem enables us to dispel darkness (Nescience or Sciolism) to apply valid knowledge for human benefits. His amazing masterpiece (Tattvachintāmani) deals with all important aspects of Neo-Nyayology as Set Theory as well. His Neo-logic Theory of legitimate knowledge is based on absolute or infallible cognition of excellence under causal condition. His style of interpretation goes through technical and terse language with conceptual skill to investigate or adjudge the given riddles in precision.

In a new aeon, the oriental nyāyika system has ushered an innovative shift to perceive legitimate source of knowledge (pramāna). Gem-brain-Neo-logician Acharya Gangeshopadhaya (Gangesh) of Mithila penned brain-gem essence of universal existence (Tattvachintāmani) which deals with all vital canons of Nyaya-Vaisheshic philosophy.

Further the work of the 1st mere treatise systematized all Nyāyic concepts into 04 main components:-

Perception

Inference

Analogy

Verbal Testimony for valid knowledge

Such 04 tools for valid knowledge are seconded by oriental logicians-Udayanācharya, Pakshadhar Mishra, Vasudeo Sārvabhaum, Raghunath Shiromani, Gadadhar Bhattacharya and the like. His Neo-Logic Theory of legitimate knowledge is based on perfect cognition of pre-eminence under causal sequence.

The 1st perception – Such cognition occurs between sense faculty and substance void of another cognition as its immediate – instrumental cause through contact component.

The 2nd (Inference), the 3rd (Analogy) and the 4th (Verbal testimony) bank on immediately previous cognition to trigger their functioning. Such intermediate feature of causal chain terminating into perceptual cognition is of 05 types as oriental logician Udyotakar opines:-

(a) Contiguity (Sannikarsha) – here one perceives not only substance but its properties and absence and so on.

(b) Contact (Samyoga) – between sense faculty and object.

(c) Companio-unified (Samveta) – The contact instantiated between sense faculty and universal or abstract concept which is represented in a concrete or visible substance.

(d) Conjoined (Samyukta) – Sense-organ and substance brought together.

(e) Inherence (Samvāya) - recognizable by essential property having something by nature of its entity–Comet shape (inherent property of Telescope using parabolic mirrors.

(f) Qualifier and qualified contact (Vishekhya Visheshanbhāva Sambandha)

According to NN Gangesh, such perception is of 02 types, typically:-

(i) Determinate (Savikalpaka)

(ii) Indeterminate (Nirvikalpaka) – reasoning in multi-Robot learning.

The 2nd –Inference- whenever one cannot determine or ascertain fact or truth of knowledge through perception, one has to resort to even other sources of knowledge i.e. Inference to confirm certain facts.

People all over the world recognize "Inference" as a valid means of knowledge and use it to ascertain and establish the substances which are beyond perception in all aspects of life. Without Inference, recognition of perception cannot be ascertained to have validity intrinsically by the integrity of causes by itself to dispel dubiety.

Further Inference is to derive logical conclusion from premises known or assumed to be true via syntax analysis on cognitive psychology. Its methodology is based on evidence and reasoning which we use data to determine observable phenomena. Reasonable Inference drawn from circumstantial evidence may properly back up findings.

Inference is vouchsafed by NN Gangesh in the Anumānkānd of his monumental work "Tattvachintamani" . Here all the 04 tools (components) alluded to above dealing with neo-logical concept prone commentary —

(i) (Anuman Gādādhari) of PL (Post logician) Gadādhar Bhattacharya,

(ii) "Didhiti" commentary of PL (Post logician) Raghunath Siromani,

(iii) "Didhiti" sub-commentary by Jagadhish Tarakalankār on "Tattvachintāmani" of NN Gangesh – are to be discussed elaborately with the Atheistic philosophers for AL –I (Artificial Intelligence) and JL-I (Judicial Intelligence) applications. His (NN Gangesh) analytical and algebraic style of exposition of technical terms has been tried straightforwardly as possible. So such version of rare contents would acknowledge the mechanism of the knowledge-base in its utmost benefit.

Almost all human beings simulate Inference in one way or the other in daily life to ascertain things or events of Non-Bio-Vision.

Neo-logician or Navya Naiyāyic Acharya Gangesh (Gangeshopadhyaya) in his masterpiece "Tattvachintamani" has devised an adroit and astute mechanism of expression which puzzles or rattles even the bonafide mentees of oriental philosophy. However I try to render the ipsi dixit of NN Gangesh intelligible with functional mechanism to modern mind. Although such trial is beset with irksome mission. So I but meet expectation that the alumni of the global gamut would treat my pitfalls which have been incidental against my pioneer efforts as a pragmatic approach to mass advantage.

Inference and its tools

The oriental logicians are one of the earliest in the planet to entertain Anumāna as a typical means of valid knowledge in Inductive Logic.

While Rishi Gautam (Nyaya) and Kanāda (Vaisheshik) upheld Inference as one of the 04 means of valid knowledge. Thereafter Neo-logician-cum-seeder –Udayanacharya and his exponents dittoed Anumān as an essential means of valid knowledge.

Later on NN Gangeshopadhyaya shot Anumāna drive through the Anuman section in his vital work "Tattvachintamani".

Such Anuman Theory mechanized around subtle or deep probe which relies on evidence and clues. It is processed through reasoning, conjunction and deduction from extra-special to special via Universal truth connecting correlation of induction and deduction under given proposition.

As a fault-free device it is the cognizant observation to get at logical conclusion. Further it refers to the process of programming adroit ML (Machine Learning) and Algorithm to deduct for future events which can be adjudged or tested within the natural flow or track of substances – valid or invalid. Firstly as an orientalist I should discuss the 04-band of tools of Inference categorically.

Now the debate is heated on the Neo-logical concept of NN Gadadhar Bhattacharya with commentary on Tattvachintamani "by NN Gangesh Upadhyaya and Didhiti" commentary of Raghunath Shiromani under impugned argument of the Indo-Atheistic-Philosophers. I have tried to leverage nut-bolts of Inference in an artistic manner:-

Take for –

(1) The command of Inference

Before defining the distinctive domain of Inference I want to decode its pitmen for its momentum in human accessories –

(a) Upādhi (Special or extrinsic pre-condition or cause for a general effect) – The utility of dubiety rests on its being an exceptional spur to analyze and evaluate its significance. Renowned paraphraser Jagadhish Tarkalankār and Reghunath Shiromani define Upadhi being exceptional cause giving over its characteristic reflection to other closely abiding its substance viz-

(1) Hibiscus flower (Upadhi) cast its redder characteristic on the quartz or crystal stone in its close contact (junction) causing redder reflection avoiding its natural reflection (whiteness). Consequently an illusive knowledge of redness in the crystal stone is apparent and not inherent.

(2) It acts on the conjunction of evidence. Further, it acts as a pervader of reason of the incidence which is qualified by some extrinsic property for Inference.

Let me treat popular example:

"Where there is smoke, there is fire"

Here the special condition as proximity crops up due to the junctional property of soaking fuel. There is not the legitimacy of smoke in the fire unless the fire caused with wet or damp wood as-

Where there is smoke, there is wet wooden cause of fire

Or

Where there is such like fire, there is smoke – logical

Here the conjunction with soaking fuel is the pervader of that which is qualified by smokehood and it is the non-pervader of that which is qualified by firehood.

Whereas I may frame syllogism-

Where there is fire in red-hot iron rod, there is no smoke.

Because the red-hot iron rod has no Upadhi – as a make-shift of the premises (Sādhya). Here such Upadhi as soaking fuel does not pervade the reason of firehood. So Upadhi is to specify counter-thesis as it has no conjunction of soaking or wet fuel. So the conjunction of a wet or soaking fuel argues the nature of pre-condition in its clear view. This conjunction of wet fuel is the determinating – concomitant (Vyāpaka) of the probandum (fire) and not the probans (smoke). Since the presence of smoke is found to be invariably attended with that of the conjunction of the soaking fuel.

In the event of the existence of the extrinsic pre-condition, the invariable concomitance (Vyapti) between the probans and the probandum ceases to be intrinsic in character. In other way, the concomitance of the probans with the probandum is instantiated itself due to the causal connection of the condition with the probans which explores its (probans) appearing as coincided with the conjunction of invariable concomitance via the probans.

But the case with reference to the fire (probandum) is not iron hot rod to be otherwise. For in a hot iron rod the presence of fire is a universal fact. Whereas the conjunction of a wet fuel is absent evidently. So it is not concomitant with fire.

Here confirmation of invariable concomitance of the fire with the smoke is rendered possible by the observation of the co-conjunction between them (wet fuel and fire). It is certified by the symptom of experience that the observation of the compresence of 02 terms cannot admit inerrancy of the concomitance.

So, as to the inferential judgement as above, the invariable concomitance (Vyapti) between smoke and fire is conditional it its nature. Lastly the causal or external or accidental as conditional property warrants the extrinsic or intrinsic character between the probans and the probandum contrived upon "Upadhi" connotation.

(B) Dubiety (Sanshaya-counter argument)-

The Nyāyic version to dubiety provides some salient principles-

At psychological attitude dubiety appears to be conscious and unconscious faith. It neither claims nor disclaims whatever things but only poses a complication for consideration. Being the such case, dubiety should be distinct from the bare lack of faith.

In dubiety, I argue more options for the same substance. According to Nyaya – Vaisheshic, a specific dubiety is undeniably an antithetic decision. Vaisheshikas comment on the affirmation or non-affirmation of dubiety as optionals and debatable as well within the premises.

But Naiyayikas comment that dubiety is indecision to ascertain any event as to tentative or weak cross examination of any event. Dubiety may be verifiable or fallacious.

In dubiety, the pendulum of reasoning between various propositions cause non-intentive correlation in fact. It has hassle of avowal due to pre-conditional cause. When I see a thick-tall object in a paddy field, I may observe to be a tall person. Here the presumption of dubiety becomes more feasible than the other. Here the indeterminate cognition is uncommon. And common property entails the cause of dubiety. That is why NN Gangeshopadhyaya talks of 02 sources of dubiety –

(1) Scepticism of Upadhi

(2) Regular cognition or perception of property common to a lot of object void of conjunction of extricating feature.

Dubiety is the desideratum of inferential logic as one wishes to prove or demonstrate or accomplish or infer optionally. Whereas (a) oriental logician Udayanacharya defines – the aspiration or impulse for syllogistic reasoning.

(b) Oriental logician Pakshadhar Mishra defines- aspiration to generate Inference in a given premises. Here it is pertinent to escape from the infinite probability of dubiety urged by Shree Harsha (Nyaya critic) with NN Gangeshopadhyaya's hypothesis of reasoning-psychological evidence on the plea of:-

(i) Reasoning qua instrument of rebuttal

(ii) Reasoning qua organ of induction

(iii) Cognizance or apprehension of concomitance to assess the requisite and universal relation between 02 locutions.

Further as the reasoning is based on invariable subconcomitance confers finity of doubt. Thereby deletion of doubt does disqualify one of the terms and deduces patent option taken for the proven fact.

At intervals dubiety may be clashing (tall stake and human) as well as contrasting (tall stake and not tall stake) terms. In dubiety there is the motion of such clashing options but no confirmed cognition of any attribute to opt between the two. So it may be called as indeterminative cognition. Besides it serves as a pro-specific or irrefutable condition of cognition of conjointly individual nature in the same event and at every moment due to the mutational motion in the mind. Such process of swinging hunch or flow of mind entails counter-argument. Dubiety seems to be the presumption of distinguishing feature of an object contrary to pro-distinctive impression of ideas.

Dubiety is not a damaging effect to one's fidelity nor one's credence (pratyaya). It serves as a campaigner for insight intelligence. It is a process to inquire into the surface significance or appreciation or estimation of an event.

Further it is not a straight way rebuttal before a rational or legitimate consent or recognition to secure maxim. Doubtful notion is a device to attend to the cogency of the given event and adjudge the consequence of the problematic puzzle. The Epistemophilic or true-knowledge seekers undergo the pitch of dubiety as a positive effort to get at the universal cognition.

Further dubiety is the process of self-reliability to enable or rectify one's own beliefs or convictions or ideology.

Although dubiety is not always logical. But the abstract reasoning is instrumental to determine genuine dubiety to evade conflicts for presumptive or speculative or conjectural intent.

In order to regulate the truth by virtue of putting forth logical justification to solve the counter-action of conflicting forces in the same given event, Neo-logic seeder Udayanacharya postulated 5-fold reasoning-devices as Modus Vivendi in "Atmatattvaviveka"-

(a) self-sufficient

(b) conjunctive or interdependent

(c) domino effect (cause-effect relationship) or cause nexus or vicious cycle.

(d) malign or faulty decisive or inevitable argument.

(e) infinity and the flexible de-infinitism.

Here NN Gangesh's hypothesis of reasoning has been arranged to be seconded by Udayanacharya as well.

The up-date debating genre on the topic makes it obvious that a conjecture (oohaa) or speculation or suspicion fails to be justified against perceptual experience as one is incompatible to the other (perception) regarding their essential feature. So the affirmative argument of conjecture by virtue of perceptual observation cannot convey conviction.

Here the counter-charge of infinitism is not so challenging as it appears. It is an avowed fact that perceptual cognition fails to justify mere conjecture for reasons alleged above. The hypothesis as such goes through name, class-character and so forth of fire in question. And it is speculative knowledge in its nature. A speculative knowledge refers to these facets of substances. In the sense of hence it can be the qualifier of the speculation of fire. A perceptual cognition of an object explores and examines the thing-in-itself and as such as indeterminate in its innate nature.

In the next step it turns to be linked with class-character, class-classification, qualitative classification, name and function. They are the entities of speculative knowledge which evidently determines cognition.

The indeterminate-perceptual cognition is the pre-condition for the access of the ensuing-determinate-speculative cognition.

Now Atheistic counter-argument against the justification of Inference may be framed as follows—

The conditional concomitance between the probans and the probandum cannot serve as the legitimacy of Inference. In inferential judgement the absence of the causal condition must be ascertained with universal certitude.

It is a pre-condition and non-compliance with it – will be curbing to the emergence of valid knowledge. There are conditions which are responsive to perception. And as such their non-dubiety may ascertain their absence. Since it stands to reason that the non-dubiety of discernible substance asserts its absence. But it is undeniable that there are conditions which are beyond one's ken of perception. So their non-dubiety cannot certify their non-presence. Certainly the dubiety of their existence is beyond to occur in the mind. And this will entail a dubiety regarding the unconditionality of the concomitance between the probans and the probandum. It is needless to discern apparently that even the far off probability of pre-conditions is malefic to the acme of inferential judgement. Although-truly the knowledge of smoke persuades even the laymen to proceed towards with the expectation of fetching fire. But this activity cannot be regarded as reliable concomitance between 02 terms. This activity owes its motive to the mere chance of the presence of fire.

The causal concomitance between 02 terms cannot be considered invariable. The effective acme of pursuit in many cases is caused by a pure presumption. Such situation deserves illustration – a person looks for fire. He discerns smoke issuing from the remote place. Here fire remains unperceived. He proceeds on the conjecture that there may be fire. He reaches his destination and finds fire. Now it is evident that this mere conjecture has been validated by the evidence of perception. And it is due to such validity that appears an Analogy or similitude of legitimacy even in a good presumption.

Now a pertinent poser pops up- An assay of the indispensable nature of conjecture makes it clear that it is not responsible to the authenticity by means of perceptual cognition. Since confirmation of one cognition by another – uniformly involves Analogy of their contents. But presumption as such refers to fictional fabrication.

In the case under review, conjecture corroborates fire – as possessed of fire universal. Here the correlation between the fire and fire – universal corresponds to substantive-adjectival model i.e. the former is the substantive in which the latter inheres as its adjective. This link of the fire with its class-character is a mere speculative framework. The universal is a pure ideal fiction. And it is devoid of extramental existence.

But the perception as a tool of cognition differs wholly from mere presumption having essential aspects.

So the Atheistic philosopher claims that perception takes into cognizance only the thing-in-itself. An entity which is unique, unassociated and self characterized falls under the need of perceptual cognition.

The perceptual cognition as such manoeuvres clear of the sphere of the class-character quality, substance, action and denomination. They are the objects of conceptual or theoretical knowledge meant for mistake. As class-character is devoid of authenticity. And as such conceptual or theoretical knowledge is not entitled to claim authencity by its own intrinsic merit. It is the perceptual experience alone is endowed with legitimacy.

The core of the debate is that the hypothesis of fire due to its speculative feature is rightly argued or certified by a speculative cognition based on indeterminate perception. The fact accounts for the defence of justification of the right conjecture of fire by perceptual cognition and the resultant presentation of Analogy of legitimacy relies on presumption. Thus the counter-charge of the Atheistic philosophers advocating only perception as a tool of cognition (knowledge) stands infallible under the demur of the defendant (inference as an organ of cognition) logicians.

But the fallibility of inference causes a severe sequel. Because the verbal evidence or testimony swings on Inference qua an organ of cognition. Since the knowledge of the locution and its key-sense is integral in a verbal testimony. The knowledge which the certain locution is loaded with a certain sense – is conditional to a series of Inference: so it is clear that the verbal testimony holds fast on Inference. Therefore I may come to tangible reasoning that the fallibility of Inference devaluates the probative or evidential value of verbal testimony.

It is the self-same case with Analogy which has been retained by the logician as stands on a verbal premises argument commented by a person having command on the subject. So an Analogy also obviously requires the knowledge of the link between the word and its sense. Eventually the fallibility of Inference is squarely malefic to verbal testimony and Analogy alike.

It is owing to this argument the atheistic philosophers have countered the sweeping consideration that only perception deserves to be an organ of cognition. Extra-organs of cognition plus inference are void of legitimacy of their own and as such liable to be deleted without any further casuistry or amphiboly.

Such challenging allegation or dispute or argument of the atheistic philosophers against the infallibility of Inference provides a perceptible or sensible (palpable) or evidential example of casuistry or ambiguity or fallacy. The upshot of their aberrant or spacious or odd argument is that the negation of infallible conditions of explicit character decries a cognitive organ of its probatory merit. Hence it is true that an organ of cognition to be considered thereby – is intent to abide by the infallible condition of explicit feature which accords a matter of legitimacy.

Now the infallible condition of explicit feature is evident by its negation of Inference. So there stands harmony between spacious cognitive organ and Inference until the negation of infallible condition of exclusive feature is considered on judging the event.

Further it meets with hitherto argument that the atheistic philosophers tilt upon the negation of infallible condition of explicit genre as the locus or hub-ipse dixit – of fallible Inference. But such plea or notion is constrained to be lethal to the atheistic attitude discussed below point-wise –

(1) The atheistic philosophers take on the exercise of negating the infallibility of Inference as a tool of cognition. But they have ignored the sequel of their emphatic postulation. They have formulated the negation of infallible condition of specific genre as the cause to beat the infallibility of Inference. But such cause serves as tantamount to the middle term of a syllogistic Inference. Eventually the cause i.e. the negation of the infallible condition of explicit genre will pitch for the Inference of the Major term viz. the fallibility in the Minor Term viz. Inference as a tool of cognition. Here the crux of contention swings between the infallibility and/or fallibility of Inference.

(2) The atheistic philosophers have evoked a device – the negation of infallible condition of explicit genre is to pinpoint their positions. But such plea is fatal to their dispute. Their argument or intercession stands transposed to a veritable Middle Term only for their sarcasm. Further their antithesis to the inception of Inference is bearable to the skeptical advocacy ignoring the infallibility of Inference.

(3) But how the atheistic philosophers can be adroit or adept with the inner cyclical phenomena. Since nescience, seepticism, pervasion and such other mental condition cannot be directly cognized by the sense-organs of the average laymen. Besides they cannot look to other exclusive type of logical instrument. Since they do not concede to the infallibility of any other logical instrument other than that of perception. They may adjudge – Inference is not an infallible logical instrument to the sapient without prior affirmation of their nescience, skepticism and pervasion about the infallibility of inference. But the case being so, they (atheists) will make themselves liable to be deemed as fanatic or decontroller

of their persuasion. So such interpretation is unbeatable as to the sceptic sort of cyclical phenomena of other identity which can only be inferred on the plea of their elocution or eloquence.

Therefore, the findings transpire- Inference is not a legitimate tool of cognition- cringes upon atheistic philosophers with tremendous effect. It drives the atheistic assent to the infallibility of Inference being adverse or countering to their (atheists) Testament. Still they cannot forego casting into their lurching arguments. The commission of non-observation of cyclical phenomena forsakes the cause of existing Inference. Since it has been already diluted upon that human attitude or conduct can be feasible on the premises of an authentic presumption. So the proclamation of the affirmation of findings – Inference which is not a legitimate tool of cognition–cannot be in anyway conflicting with the pioneer position of the atheistic philosophers. So it is obvious or self-evident that they recognize only the legitimacy of perception as a tool of cognition. And as a result of which they proscribe the concurrence of the infallibility of inference as its tool of cognition. Because they have rebuffed verbal testament as the conditional tool of cognition themselves.

If adversely they feed the fallibility of findings "inference is not a legitimate tool of cognition" – this would sum up to the assent of the infallibility of inference.

(4) Since the fallibility of Inference is not legitimate tool of cognition implies that it causes fallacious verbal testimony. And emerges on a micro or nanoscopic grain of truth that the fallacy in the ensuing case lies in the cognition of the negation or non-existence of the infallibility on the condition of the existence of the negation of the negation of infallibility. If the inferential judgement is not a legitimate tool of cognition-should be a fallacious judgement. Consequently, such conditional proposition would entail the existence of the negation (true) of the negation (false) of infallibility in the logical subject or data-base viz. inference and the succeeding phenomenon of the cognition of negation of infallibility in the logical subject. Here it crops up as the cognition of the negation of negation of infallibility. On the condition of the existence of the negation of negation of fallibility would be a legitimate data logically. But in the presumption under critical analysis, there is the negation of infallibility in Inference. And notwithstanding the datum that the cognition of the negation of infallibility has been materialized regarding Inference. And such

locus has provided the cognition of the negation of infallibility in inference – negative fallacy. Here the atheistic favouritism or mind set or opinion is prone to blatant contravention or confliction.

Such negation of the negation of infallibility sums up the infallibility itself as the Law of Negation in Mathematics or Logic which confers positive computation. To be more clear –02 negatives [(-) × (-) = +] are invalidated for its validity. Again if the statement is false, its negation is true (legitimate) and Vice Versa. So the recognition of the negation of fallibility in Inference corresponds to the recognition of its infallibility. And apparently it calls for confliction or impugnation in terms so far as the atheistic conflict is involved.

(5) The atheistic philosophers look into a new defending logical device- they assume that the presentation of fallacious findings discards the legitimate benchmark or touchstone of fallibility. Since fallibility may be evolved owing to the negation of legitimacy of the tool of cognition.

It relates to the comment on the legitimacy of being a tool of cognition and its negation that are efficient to debar or rule out the feasibility of the 3rd option or substitute or experimental. To amplify-the act of being a tool of cognition calls for the negation of the authenticity of being a non-organ of cognition and the otherway inversely.

The reliability or authenticity of being a tool of fallacious findings and that of being a tool of infallible cognition are not commonly absolute or exclusive. There lacks the least consent to follow that which cannot be a tool of legitimate cognition – is constantly a tool of fallacious findings and the otherway inversely.

Now as per atheistic attitude, the hypothesis –

"Inference is not a legitimate – instrumental device of cognition" – is neither effective of fallacious findings nor of a legitimate device.

On the other hand, it brings about a 3rd kind of cognition which is exclusive or uncommon or an independent criterion in its nature or temper.

Actually the terms being the essential proposition cause an indeterminate judgement. However, subsequently they act as the medium to transmit or communicate or provide connotative sense. And thereby they emerge as instrumental to the emergence of a determinate psychic (mental) cognition. So it is clear that propositions as such are not immediate condition for the emergence of the exclusive genre of cognition as referred to above.

The purveyance of connotation provides as an instrumental regarding the emergence or evolution or realization of the determinate psychogenic (mental) cognition. I mean to say that it is immediately anteceded by the purveyance of connotation derived from the propositions. A determinate cognition is invariably or evenly merged by a qualifier (adjective). To analyze the fact, the propositional terms match up or conform with the condition of qualifier (adjective) and eventually evolve into dynamic effect of the determinate psycho-cognition resulting the terms being the immediate condition of its emergence. Such sort of determinate psycho-cognition is unlike verbal testimony caused by its exclusive link to the cerebral faculty. So the determinate psycho cognition is operative to engender rational or logical enterprise conceivable. Therefore, the premises, "inference is an illegitimate tool of cognition" could be taken on to convey an exclusive concept. So the allegation of contravention against the atheistic philosophers posits quite refutable or challengeable.

However, the above verbal debate or dispute of the antagonists or adverse challengers is devoid of probatic merit. The fallibility of inference qua a tool of cognition would altogether counteract or foil the perception alone being the infallible instrument of cognition defended by the atheistic mind.

Because the infallibility of cognitive tool is attested or accredited by Inference. The physical property of infallibility per se (as such) is imperceptible or mystical or paranormal or extrasensory and immune to perceptual cognition.

It is the Inference in particular is just qualified and chartered to the infallibility of a given tool of cognition. So the rebuttal or rejection or retraction of the infallibility of Inference would contribute to the assertion or proclamation of argument or perception alone is undoubtedly problematic and incredible.

Further it cannot be argued that the infallibility is metaconscious or self-perceptive in its key-nature. And the privilege of Inference can be waived of its presumption. Since if the infallibility be held as self-perceptive, this would entail the emergence of dubiety implausible. So the cognition of the patent position causes the recognition of the infallibility of Inference in terms of logical compulsion or need.

(6) Nevertheless the atheistic philosophers do not quitclaim on their basic argument. Further they counter that the claim – "Self-perceptive or meta consciousness would ban the feasibility of dubiety – is mistaken or fallacious. Since the causal condition to cause dubiety on infallibility is apparently off to the perception. It assumes to be exaggeration to remark essential feature or character that affords the justified reason for fallibility of a given tool of cognition. But as commented earlier that – "perception per se relates to the bare truth-that is – the essential character or feature-in-itself. This presents the vast bed-position to common feature of its character reference. So the negation of the term or pre-requisite of dubiety close to infallibility regarding perception qua a tool of cognition provides the phenomenon of dubiety about its infallibility implausible. But for the perception is causal with the essential character-in-itself which is exclusive and factual in its innate nature.

The phenomenal concept of perception rules out the likelihood of dubiety. Since dubiety is an illusory or make-believed objective reference. Such rejection from the range of dubiety certifies the negation of fallibility in perception qua a tool of cognition. Moreover I have discussed about the infallibility of perceptual cognition being self-perceptive or meta-conscious or spontaneous. So the negation of the terms of dubiety regarding infallibility of perception as a tool of cognition defends the counter-argument of its innate self-legitimacy.

(7) Further the another counter-argument is hit with reference to valid fact that calls upon a given tool of cognition to claim its exemption from the trespass on fallibility. So the phenomenal fact is the only benchmark of infallibility regarding a tool of cognition. Further as indeterminate perception devoid of subject-predicate link which follows this infallible condition undeniably. The phenomenon fact of this case-event rules out or debars the doubtful option of its fallibility long way off.

The syllogists are skeptical or sceptical to the conviction or infallibility of such counter-argument. They argue that the dubiety-whether a given perceptual cognition is infallible or not flows into the mind in usual condition. They shoulder their responsibility to furnish a logical rejoinder.

The atheistic philosophers argue that the dubiety of the syllogist (proponent) hinges on the confusion or misinterpretation or misconception of the entire locus. The dubiety about legitimacy or infallibility crops up in all intents with regard to the determinate perception – attributive knowledge substantively.

There is a chain linking the phenomena of 02 kinds of perception referred to above-

(1) Such is the indeterminate perception that directly preexists or follows the determinate one.

(2) And due to this nearest or adjacent sequence, the dubiety appears or crops up for the latter – is hypothetically accredited to the former. So the challenge of the syllogist that self-legitimacy of a tool of cognition would hinder the chance of dubiety – suffers rejection in toto. And that being so, the self-legitimacy of indeterminate perception debars probe on the frail or unstable or delicate or feeble footing of the futility of dubiety.

However, such counter-argument of the atheistic philosophers is estimated or pre-planned to veil the innate fault or flaw of his standpoint. They have come to a bootless errand to elude or outwit the essential topic connected with the existing debate – the characteristic of legitimacy is an extrasensory or preternatural or transcendental in its innate character. It stands away from the panorama of perceptual cognition. They (atheists) insist on depending the self-legitimacy of indeterminate perception in utter impractical obsession. It is but reasonable that the extrasensory nature of the legitimacy can only be perceived or cognized by virtue of Inference. Whereas the counter-argument that the indeterminate perceptual cognition or noesis cognizes itself with its innate legitimacy simultaneously – is the only emphatic or arbitrary or adamant allegation. Thus, the objection of the distinct or separate or unique parameter of Inference as a tool of cognition or noesis is committed to deter the persistence or tenacity of legitimacy of perceptual cognition all square.

As the cognition refers to mental strategies which catalyze intuitional memory, reasoning, perception and learning skill to obtain knowledge. I mean to say that the cognition itself is inferable drawn as conclusion from prior knowledge and present evidence. Because the cognition presupposes some other cognition. On the nerve of pervasion or invariable concomitance, the cognition of Ghata (pitcher) reveals the physical object or entity "Ghata" sharing 02 characters:-

(a) State

(b) Demeanor (mien, look, attitude and appearance)

The probans (cause) of a cognition is the essential character or property of being cognized (gyātatā), when an object such a "Ghata" is made known and I come to know the "Ghata". So the intrinsic character of the property of being cognized is the probans (reason) by virtue of which we infer cognition.

Further the property or essential character of being cognized (gyātatā) differs from that of being the object (Vishayatā) of the Neo-Logical system interpreting it in terms of the characteristic relation between cognition and its object. Here the cognition of the probans (reason) is needed prior to the inferential cognition. If the cognition of the probans is conditional on that of the probandum, I cannot infer the probandum (fire) which is a new cognition and none is capable to entertain the relative property.

On deep and delicate decoding for determination of the momentum of Inference, Neo-logicians have enkindled their intelligence to defend the absolute status of Inference Analogy and verbal testimony as tools of cognition. They posit that there are 02 underlying conditions in the affairs of inference-

The assertion of the relation of pervasion (Vyapti) between the probans and the probandum as Middle and Major terms respectively.

The presence of the probans in the knowledge –base as minor term.

Thus the inevitable significance of the term "Inference" is well secured as instrumental to condition.

The definition and nature of Inference

I proceed to deal with the definition and nature of Inference qua an instrument of cognition. It is remarkable to note that cognition as such is an act and is conceived by an instrument. It is for the performance of the act that an instrument is pressed into service. Even a slight alteration in the nature of an act entails a corresponding alteration in the nature of an instrument. Further an instrument is bound to vary with the variation of an act. The fact is too obvious to elaborate. So a true determination of the nature of resultant act is essential for the comprehension of an instrument. It is the bare fact that inferential cognition is an act and as such its prior determination is vital for the comprehension of its instrument. Due to this reason NN (Navya Naigāyic) Gangeshopadhyaya (Gangesh) has set forth the definition of the inferential cognition despite the fact that his main pursuit is to formulate the definition of Inference qua an instrument of cognition. He has propounded the definition of Inference as an instrument of cognition subsequent to the formation of inferential cognition.

Besides the determination of an instrument of cognition involves Inference to cognition. So a right comprehension of the nature of cognition is indispensable for the formulation of the definition of an Instrument of cognition. It is owing to this fact that a thorough grasp of inferential cognition is needed for the accurate definition of its being instrument.

This consideration has persuaded the author of Nyaya "Tattvachintamani" to formulate the definition of inferential cognition by leaving aside the case-instrument to be defined later on.

Now a critical analysis of the definition of inferential cognition as propounded by NN Gangesh brings to light its essential conditions–

1. The knowledge of invariable concomitance between the probans and the probandum.
2. The existence of the probans in the logical subject or knowledge-base (pakshatā).

The knowledge of these 02 factors is conditio sine qua non of all species of inferential cognition.

It deserves to be mentioned that the knowledge of invariable concomitance (Vyāpti) must synchronize with that of the existence of the probans in the logical subject or knowledge-base. To be more precise the knowledge of these 02 conditions must maintain co-evality (coincidence) in the matter of their occurrence. The compliance of this condition is indispensable of the materialization of inferential cognition.

Now I propose to consider elaborately the nature of invariable concomitance which posits as pivotal in the matter of the culmination of inferential judgement.

A section of logicians avers that the invariable concomitance implies the non-existence of the probans in locus characterized by the negation of the probandum. It is evident that they have viewed the relation of invariable concomitance in its negative aspect. A concrete instance is set forth in strict syllogistic form which renders the position clear –

The hill has fire – probandum

Because it has smoke – probans

I should explain the technical terms about "probans" and probandum in a lucid way as –

(1) Probans – is a fact given in evidence as a proof (evident fact) of another fact.

(2) Probandum – is a fact that requires to be proved (evidentiary fact) for defence to logical conclusion.

Now the present instance, the probans viz. smoke stands in the relation of invariable concomitance with the probandum (fire).

Since in a lake or damp land, there is the negation of probandum (fire). And the non-existence of the probans (smoke) in them stands confirmed by the verdict of experience. This non-existence of smoke in a locus characterized by the negation of fire vouchsafes the existence of the relation of invariable concomitance between them.

Let me discuss about the relevance of the issue

Now the definition under review-the expression-negation of fire is the counter-positive of the negation. Here counter-positive is an identity whose existence is negated. The intrinsic character of counter-positive of negation of such must have reference to its determinant in the shape of relation and essential attribute. Here the total entity is supersubstratum of its micro-parts as earthen pot or cloth in inherent relationship. In the event of the absence of express mention of the relation and essential attribute as the determinant (Avachhedaka Sambandha) of counter-positiveness, it may be taken in an arbitrary way. Further Ghata or Bhatagrah (earthen water jar or pitcher) is substratum and Ghatatva (pitcherness) is – supersubstratum. When I utter the word as "Ghata", I undoubtetly recognize the "Ghata" substance and not of its size, shape, colour or the like due to the inherent relation between substratum (Ghata) and supersubstratum (Ghatatva vanpitcherness) in syllogistic maxim. This Ghatatvavān (super substratum) lies in the "Ghata" by correlation. Substratum and supersubstratum are identified by each other. In logical terms I may connote them as location – locative link (Dharmadharmibhāva Sambandha or Adharadhey Bhava Sambandha). Here the locus of "Ghatatvam" is "Ghata". Clearly the pitcher is location and pitcherness is locative or dharma. Such relation of qualificand (pitcher) and qualifier (pitcherness) is universally inherent in Neo-logical term.

Here the pitcherness is the physical property or characteristic of 'pitcher'. Because it maintains its intrinsic or composite property unchanging in the compound (pitcher) rather changes in shape, size of the pitcher. Such is the case with a yarn or its cloth. Here the materials (muds etc) involved is inherently the same before and after physical changes. It may change into the physical state of the substance without affecting chemical composition. The substance may be distorted in shape and size and not in essential or fundamental property. The mud may not be dissolved into chemical one. The pitcher is made due to ductile character of the mud (earth) through many physical changes intensively or non-extensively.

Here pitcherness (Ghatatva) is counter-positive (pratighatmaka or pratiyogi) of the pitcher and the pitcher is the base-relation by contact relationship in a responsible mode. And the relation of contact renders the counter-positveness (Pratiyogita) of the negation of pitcherness in the pitcher.

In other words, the pitcherness is absent in the pitcher by contact relationship but it presents in the pitcher by the relationship of inherence.

This counter-positiveness of the negation of pitcherness as determined by the relation of contact is the Neo-logical interpretation.

Here negation has one counter probability and one locus. So "pitcher" has base-relatum (Anuyogi) & pitcherness has counter positive relatum (pratiyogi) of negation.

Here the Neo-Logic transpires the relation between property or essential character (Ghatatva) and property holder (Ghata). Further counter relatum is a pitcher and base-relatum is a pitcher-maker. This Ghatatvam leads to non-ambivalent sense in the "Ghata". Here the counter-positive propositional instance –

If not 'A' then not 'B' and inversely

if 'A' then 'B'

Here the propositions are formed by contradicting both the subject and predicate or both – hypothesis and conclusion interchanging them.

It is in the fitness of thing to note in the Neo-logical parameter that there are 04 kinds of relation mainly-

(1) Contact (Samyoga)

(2) Inherent (Samavāya)

(3) Natural State or Character (Swaroop)

(4) Identical (Abheda)

Besides there are many concepts of relation in Neo-Logic. But I should analyze the Law of Causation (Kāryakarānbhāva) between 02 relata-

(1) pitcher – maker

(2) pitcher on cause and effect relationship

Further the relation of 02 properties (cause-object and effect-motive) in 02 different relata is termed as causeness (karanta) and effectiveness-karyata).

In other way, the relation may be of 02 types in Neo-logic –

1. Occasion-determining (Vrittiniyāmaka)

   Where a phenomenon in another phenomenon – as the fire can exist on the hill confirming relation of contact (Samyoga) between fire and the hill.

(2) Non-occasion determining (Vrittyaniyamaka)– where a phenomenon cannot exist in another phenomenon as when a person standing on the ground touches a tree with his hand confirming the relation of contact between the person and the tree. But the person does not exist on the tree appearing no relation of contact with the tree.

Let me deal the matter universally-

The universe prevails on twin phenomena-

(i) Substratum of properties (dravya or matter) as nucleus endowed with super faculty (dharmina).

(ii) Common or Universal (Sāmānya) – as supersubstratum or characteristic property.

Here the mystical relata in relation to the phenomena occur through conjunction (Samyoga) and inherence (Samvāya) on underlying phenomenon of occasion (vritti)

Further –

(a) Dravya – the pitcher –characteristic property. The reason is –super characteristic.

(b) Universal – pitcherness relates to a particular Matter (mud) as relation of inherence.

A person can locate determinant which serves as occasion determining relation in the affairs of Inference – the hill has fire.

Since it has smoke.

Here the probandum or the phenomenon whose existence is to prove– is fire and the probans or the cause (Hetu) is to prove– is smoke. So the smokehood of the smoke is the knowledge-base to confirm that hetuta (causation or the existence of cause) lies in the smoke.

The occasion undertakes that smokehood carries on the Ist condition of the determinant being the locus of the determined (probanshood) must be the locus of the determinant (smokehood). Here the determinant works as determining relation. Both determinants are needed to determine the relatum to the smoke existing through contact relationship.

Likewise, if one takes the negation of probandum (fire) in the relation of Inference, it may be the determinant of the counterpositive in this case. And the conclusion – the negation of fire in the hill would be possible on the basis of the relation of inherence. And it is needless to observe in the authentic instance of invariable concomitance between fire (probandum) and smoke (probans). The probans (smoke) will not be non-existent in the locus bereft of the probandum (fire). This will militate against the nature of invariable concomitance as alluded to above.

Therefore, the distinct type of relation serving as the determinant of counter-positiveness (fire and smoke) must be stated in the most explicit term. In the case under consideration, the relation which is the determinant of the intrinsic character of the probandum (fire) is to be understood as the relation serving as the determinant of counter-positiveness. And this logic would do away with the objection alleged above.

Since it is relation of conjunction (Samyoga) which is the determinant of the intrinsic character of the probandum (fire). And it is too apparent to be required to mention that the probandum (fire) cannot be regarded as absent in the hill on the basis of the relation of conjunction (relative cause). The explicit reference to the relation of conjunction serving as the determinant of the intrinsic character or nature of the probandum (fire) rules out the feasibility of resorting to the relation of inherence (Samavāya Sambandha) ad libitum. Now let me evaluate the result of this Amendment–

The probandum (fire) is absent in lakes on the basis of irrelation of conjunction and the probans is also non-existent in it. Here it is obvious that such edification or pedagogy of syllogism evades the charge of non-existence to which the determination of invariable concomitance stood exposed.

Thus, the amended definition of invariable concomitance may be stated as follows-

The nonexistence of the probans (smoke) in the locus characterized by the negation of the probandum (fire) in which (the probandum) counterpositiveness is determined by the relation serving as the determinant of the intrinsic character of the probandum (fire).

Broadly NN Gangesh (Navya Naiyayic Gangeshopadhyaya) agrees with pararāthānumān (for else) to deal through 05-step of syllogism to carry on conviction for others as observed by prashastapāda Bhāshya on Vaisheshic Dharshan also regarding the accessible clarification of Inference–

Overall inference is based on rational observation, prior legitimacy, Law of Causation and reasonable probe to get at the new results or truthful deduction – as observing smoke and inferring the existence of fire under penta-step-

(1) Hypothesis or knowledge-base or Induction or proposition (Pakshata, Sādhya) – The hill has fire – (Angnimān parvat).

(2) Reason or causation (Ling, Hetu) – Because it has smoke to ascertain the proposition (Dhumāt).

(3) Illustration or example (Udāharan on credibility or pervasion) – where there is smoke, there is fire (Vyapti).

(4) Reaffirmation or application (Upanaya) – The hill has smoke which is invariably associated with fire (Tathā Cha Ayam)

Vyāpti Vishishtasya Hetoh Pakshadharmatā Pratipādaka Vachanm up-nayah.

(5) Conclusion or judgement (Nigamana)

The hill has fire (Tasmānt Vahnimān parvat)

Here the serial nos. 02 to 04 are based on analogical syllogism and there may be called legitimate deductively. So the Inference is not the echo of prior data rather it formulates a patent promotion of knowledge.

Now the Inference may be of 03 kinds –

(1) Existible in advance on prior experience (Purvavata) – is based on prior experience of invariable concomitance (Vyāpti) in 02

phenomena as an access from cause to effect. Because cause or reason (Linga) is invariable and unconditional antecedent of effect and an effect is the invariable and unconditional consequent of the cause or linga. When we infer an unobserved effect (rainfall) from the observed reason (Linga) – cloud. We have purvavata viz. we observe the dark cloud in the sky and infer future rainfall from the dark cloud or would be rainfall.

(2) Residual effect of prior incidence (Sheshavata) – it is based on causation (linga) when we infer an unobserved reason from the observed effect. Further we have Sheshavata viz. when we infer prior rainfall from the rapid muddy – current in the river – as it rained nearabout.

(3) Commonly observed henceforth on contemplation remotely (Sāmānyato drishta) – It is based on non-causal constancy of co-existence- as when we infer cloven hoofs by its horns based on pervaded-pervedor (Vyāpya Vyāpak Bhāva Sambandha) as unconditional or constant concomitance on analogous footing which certifies the legitimacy of conclusion. Further we infer the Sun, the Moon, Stars to be locomotive as seen to change their location. Thus Vyāpti is the co-existence of the probans and the probandum in all the segments and tiers of the probans and does not coincide in absence between the probans and the probandum. Thus the critical and contemplative expertise enhances prior knowledge and experience to secure or identify unknown or uneducated facts. The code of Inference can steer to evaluate situations and deduce them in their integrity.

The Character of the Subject of Inference (Pakshatā)

Now I propose to deal with the problem of Pakshata or the essential character of the subject of Inference. It is one of the fundamental postulates of the Neo-Logician that Pakshata is the condition of Inference. The exposition pakshata has been operated in a technical sense and accorded a precisely delimited connotation of its own. The older logicians aver that the dubiety of existence of the probandum in the subject (Minor term) constitutes the connotation or the essential character of the subject. This recognition ultimately involves that dubiety is the condition of Inference. The Oriental logicians concur in maintaining that Inference cannot occur in the event of the antecedent knowledge of the predicate in the subject.

There is a deeper reason behind the adoption of this attitude. Inference being an independent cognitive tool is expected to deliver a novel information. And obviously the prior knowledge of the existence of the probandum in the subject is bound to defeat this objective of Inference. It is owing to this reason that the older logicians have banked upon the dubiety (indecision) of the predicate (Sādhya) as its Universal condition. This situation deserves further argument –

The prior knowledge that the subject is possessed of the predicate – eliminates the chance of dubiety qua condition of Inference which does not permits its emergence. But if the doubt precedes the knowledge of the existence of the predicate in the subject, this serves as an indicator of the absence of the Latter (prior knowledge of the predicate of the subject). The consideration of the fact has persuaded the old logicians to refrain from recognizing the prior knowledge of the probandum (fire) as an impediment to Inference. And as its ultimate consequence they have also refused to accept the absence of the prior knowledge of the probandum as the essential character of pakshata qua condition of Inference.

They are positive that the presence and absence of dubiety (suspicion) would account for the emergence and non-emergence of the inferential judgement. I have adverted to the fact that the dubiety of the probandum qua the essential character of the subject – is the condition of Inference.

The dubiety that the probandum may occur with respect to the subject which is different from the subject in which the probandum in question is actually sought to be inferred. And it is obviously needless to observe that the dubiety of the probandum cannot be the condition of Inference of the probandum in the intended subject.

To be more precise, the dubiety of the probandum must have a clear reference to the subject in which it is desired to be inferred. That is to say – the dubiety of the probandum (fire) in the woodlit-kitchen will not be the condition of its Inference in chill. It is the dubiety whether the hill has fire or not – should be regarded as the condition of the subsequent of Inference of fire in the hill.

Now a pertinent question arises –

It is the felt fact that 02 distinct types of dubiety :-

(a) whether the hill has fire or not

(b) whether the fire is existent in the hill or not – precedes the inference– the hill is possessed of fire.

The distinct character between the 02 types of dubiety is accounted for by the transposition of the 02 terms –

(a) fire (b) hill

Now if it to be asserted that only the former type of dubiety is the condition of the Inference of fire in the hill, the latter types of dubiety will be denied of its legitimate command.

In this predicament or dilemma, oriental logician Raghunath Shiromani has made the following observations – There is the subject in which the determination of the inferable predicate on the basis of a particular relation entails the cessation of dubiety about it. And this very dubiety will be the essential character of the self-same-subject in which the predicate is intended to be inferred on the basis of the self-same-relation. Here the evaluation of the results yielded by curvy or tortuous statement is essential. Now the 02 types of dubiety alluded to above is susceptible for elimination by the determination of the existence of the predicate in the subject. The dubieties- whether the hill has fire or not.

Whether the fire is existent in the hill or not – are equally removed by the determination that the hill (subject or Dharmi) is possessed of fire. And consequently these 02 varieties of dubiety will be indiscriminately entitled as the essential character of the subject qua condition of Inference – that the hill has fire. So it is evident that this amended interpretation of pakshata successfully evades the charge of non-comprehensible as alleged above. So it is obviously needless to multiply instances. All possible cases would be comprehended by the present interpretation as set forth by oriental logician Raghunath Shiromani.

The present interpretation of pakshata deserves further consideration –

Suppose, the dubiety of the predicate occurring on the basis of the relation of inherence and its subsequent determination takes place on the basis of the relation of conjunction.

Now if the Inference of predicate is intended on the basis of the relation of conjunction, the dubiety of the predicate on the basis of the relation of inherence would not be invested with the status of pakshata with reference to the Inference of the predicate in question.

Since the relation upon which the dubiety of the predicate is grounded and the relation on the basis of which the Inference of the self-same-predicate is intended must be one and identical. And this has been clearly laid down in the body of the definition of pakshata under review.

I have shown that dubiety is the condition of Inference. And analysis of the nature of dubiety is apparent that it is liable to cessation by the determination. If it be the case-why contradiction (Bādha) which is an accredited type of fallacy of inference will be denied by the rights of being considered as an essential character of the subject qua condition of Inference. Since both – dubiety and contradiction – are equally eliminated by the determination of the existence of the predicate in the subject. Both of them share commonly the characteristic of removability by determination which stands on a par with dubiety and as such deserves to be recognized as the condition of Inference. Moreover, the contradiction operates as a bar to the knowledge of the predicate in the subject. And the proponent who advocates that "dubiety is the condition of Inference is also based on his argument that it (dubiety) directly counter-acts the knowledge of the existence of the predicate in the subject. The views of advocacy by oriental logician Reghunath Shiromani may be summed up as follows –

(1) There must be a dubiety and this dubiety must be liable to be eliminated by the determination of the predicate in the subject. But this poses a serious objection –And it may be elaborated by a concrete instance.—

The dubiety may occur –

Whether a fireless hill has colour not (possessed of). Now this dubiety is liable to be resolved by the determination –"the hill has fire". And this conclusively proves its compliance with 02 fundamental conditions of pakshata viz-

(a) The existence of dubiety
(b) Its removability by the determination as referred to above.

So this dubiety will be invested with the status of pakshata with reference to the intended Inference – the hill has fire. But such should not be the case.

Since the case under review, the dubiety of the nature of contradiction – for the comprehension of the firelessness in the hill will be an impediment to the Inference of fire in it. Therefore, if the present dubiety qua contradiction (Pratibandhaka) viz.

Whether the fireless hill has colour or not be regarded as the essential character of the subject, this would entail the overextension of the interpretation of pakshata as propounded above. Since it is the universally avowed fact that contradiction acts as a prevention to the emergence of inferential judgement. And the prospect of the implementation of the interpretation of pakshata to the case of dubiety qua contradiction renders the charge of over-extension inescapable.

It may be argued that even if the present interpretation of pakshatā overlaps to the case of dubiety qua contradiction, yet no desirable consequence is likely to follow from the situation. Since dubiety qua contradiction would foil the occurrence of Inference. And this very fact would prevent the possibility of its receiving recognition as the essential character of the subject qua condition of Inference. So the over-extension of the definition of pakshata to dubiety qua contradiction appears to all intents and purposes safe and harmless.

But the critical analysis of the present case shows that this dispute is absolutely untenable. The present instance "whether the fireless hill has colour or not" – has 02 constituents parts –

(1) fireless hill implies the comprehension of firelessness of the hill and as such is the case of contradiction. Because it is bound to checkmate the materialization of Inference of fire in the hill.

(2) Whether the hill has colour or not implies dubiety of the existence of colour in the hill. Noteworthy of remark that the present instance illustrates only one and unique type of apprehension viz. dubiety qua contradiction or contradiction qua dubiety. Now the apprehension of firelessness of the hill is susceptible to the dubiety – whether it is assailed by invalidity or not.

To be more precise, this dubiety of illegitimacy occurs only with reference to this portion of instance under consideration. And this prospect of assailability (susceptible to forceful attack) of the dubiety of illegitimacy would not permit it to operate as a bar to the emergence of Inference of fire in the hill. Now if the interpretation of pakshata over-extends to this dubiety qua contradiction, this would entail a malefic consequence.

Since this dubiety qua contradiction will be invested with the status of pakshata qua condition of Inference and paves the way to the materialization of Inference. But this would hinder the fundamental thesis of the proponent. The situation deserves further clarification – it has already been alluded to above that in case of the existence of the prior knowledge (Siddhi) – "the hill has fire" – the Inference of fire in it – does not take place. It is in strict compliance with the situation that dubiety has been accorded the status of pakshata qua condition of Inference. The presence of dubiety – "whether the hill has fire or not" – the hill holds fire (a particular character). And this facilitates the emergence of the Inference of fire in the hill.

Now dubiety qua contradiction, due to its assailability of the dubiety of illegitimacy as referred to above, has been deprived of its power of operating as a prevention. And as a result of which it would fail to act as a hindrance to the prior knowledge and the prospective Inference.

The proponent has advocated that Inference cannot occur while there is the prior knowledge of the inferable predicate. But as upon the dubiety qua condition has been conferred the status of pakshata qua condition has been conferred the status of pakshata qua condition of Inference.

This would materialize Inference possible when there is the prior knowledge of probandum. And it is needless to observe that there is in direct dispute with fundamental standpoint of the proponent (Sanshaya Pakshatavādin).

The proponent counter-argues that the charge of over-extension of the definition of pakshata is counter-factual or inapposite or inaccurate. Since the expression "Dubiety" which has been implied in the body of the definition of pakshata has its own technical sense. It implies a species of cognition in which there is the absence of the attribute-of-being-subject (vishayatā) with reference to the object that operates as determinant of the attribute-of being-an impediment (prabhandhakatā) to the determination that "the subject holds the predicate". This cumbersome statement hits verbomania to be unconversant with the techniques of the Neo-Logic. So it deserves exposition – cognition as a class is the subject (Vishayi) with reference to its object.

The proponent claims or asserts that there is the distinct type of cognition which does not be the subject with reference to an object that plays the role of determinant of the fact of being an impediment. It is needless to remark that this impediment must be one which counteracts the ascertainment of the existence of the probandum in the subject. And it is the species of cognition that has been desired to be called "dubiety". Now let me revert to the instance which forms the bone of contention.

It is - "whether the fireless hill holds colour or not". It has already been indicated above that if this dubiety qua contradiction were not assailable by the dubiety of invalidity, this would frustrate Inference -"the hill is fireless (Bādha) – acts as a bar to the apprehension – the hill holds fire".

In the case under investigation, the dubiety qua contradiction has for its content – "the firelessness in hill". And as its logical consequence – it also operates as an obstruction to the possible Inference of the possession of fire with the hill. And it calls for observation that both contradiction and dubiety (Badha & Sanshaya) qua contradiction have the same effect viz. firelessness in the hill as their contents (Vishaya) and both of them act as a hindrance to the emergence of inference – "the hill holds fire".

The fact that dubiety qua contradiction assailed by the dubiety of illegitimacy is powerless to debar the Inference of fire in the hill, is a different thing. For it is the dubiety of assailability of illegitimacy that accounts for its significance.

But it is undeniable that both of them stand on a par due to the identity of their content. And the consideration of this fact reveals that the dubiety qua contradiction also has the content – firelessness in the hill which deserves to be regarded as the determinant of the attribute-of-being-and impediment to the determination that the subject –"the hill holds fire". And it clearly demonstrates that the dubiety qua contradiction is a species of cognition in which there is the attribute-of-being – a subject (Vishayatā) with reference to the object – firelessness of the hill". This very attribute of being – a subject is the determinant of the fact of being an impediment to the determination of fire in the hill even in the accredited case of pure contradiction.

The cognition which refers to the identical subject is regarded to be held of identical attribute – of being-a-subject. And a compliance with the criterion is bound to be conceded that in both cases viz.

(a) dubiety qua contradiction assailed by dubiety of illegitimacy.

(b) genuine contradiction – the attribute-of being- a subject (Vishayata)

as the determinant - of being – an impediment – are identical. Therefore, the dubiety qua contradiction is not characterized by the absence of the attribute– of being – a subject as the determinant – of being – an impediment to the determination of fire in the hill. And the net result is that the allegation of overextension and entailment of Inference even in the case of the prior knowledge of fire in the hill get completely smashed.

Oriental logician Jagadisha Tarkālankār has expressed the standpoint of the proponent (Sanshaya pakshatāvādin) in a nut shell.

It has been referred to above that Inference does not take place when there is the knowledge that the subject holds probandum.

Now what is the nature of this prior knowledge?

It is perceptual in character answered by the proponent. In the event of the presence of perceptual knowledge of the predicate, its Inference cannot occur in normal circumstances. The reason is not far to seek. –

It has been reiterated ad-nauseam that dubiety is the essential character of the subject qua condition of Inference. Now the existence of the perceptual knowledge renders the occurrence of dubiety impossible. And consequently the absence of dubiety which is tantamount to the absence of condition accounts for non-emergence of Inference. But what will be the case – if dubiety precedes the perceptual knowledge of the predicate in the subject? The rejoinder is obvious – the preceding dubiety stands the relation of opposition to the perceptual knowledge. Resulting that it would not suffer the presence of the perceptual knowledge in question. So Inference will take place without encountering any complication or trial. Granted that the antecedent dubiety debars the chance of perceptual knowledge of the probandum in the subject. But what would be the locale – if the knowledge of the existence of the predicate in the subject be the nature of recollection? It is true that dubiety counter-acts perceptual knowledge in all conscience. But recollection as a variety of knowledge is distinct from the perceptual one and as such stands on an unequal footing. Oriental logician Jagadisha Tarkalānkār has adjudged the cogitation (uhaapoh) –

It has been asserted that there is no animus between the antecedent dubiety and recollection. And consequently the former will not counteract the latter. And this inevitable results would come out that in spite of the presence of the recollection of the possession of the probandum of the subject, its inferential judgement would take place on the basis of dubiety qua condition of inference.

Certainly the proponent has been called upon to face an intriguing effect. The skepticist or scepticist (Sanshaya Pakshatavadin) categorically affirms that Inference would take place despite the existence of recollection.

But a pertinent question arises – the proponent maintains that if there be the perceptual knowledge, the emergence of Inference does not take place. But in the event of the existence of recollection which is also a species of knowledge, Inference finds a smooth and easy passage.

Further oriental logician Vātsyāyan – celebrated critic or expositor of scholiast of "Nyayasutra" has observed that perception as an accredited tool of cognition occupies the superior position in comparison with Inference and verbal testimony. Its overall supremacy over other kindred cognitive tools has been emphasized in more lucid terms by him. The expert critic has averred that a person may possess the knowledge of an object or a thing by means of verbal testimony. But in spite of that he, again, desires to confirm the information yielded by the verbal testimony resorting to the Inference.

Besides not only this, he again calls into requisition of the service of perception for the final confirmation to deliver Inference. This fact absolutely confirms that perception occupies the position of the supermost significance in the realm of cognitive tools. It serves as the final court of appeal for the verdict to which carries its indisputable command. This observation of Vatsyāyan has exerted a wise influence on the mind of the proponent who belongs to the older school of Oriental Logic.

It is owing to this reason the proponent is inclined to advocate that inference does not take place in the event of the presence of perceptual knowledge which is endowed with superior probative force.

Now when the supremacy of perception over Inference and verbal testimony is not liable to dispute, it follows as an essential porism or corollary that recollection, also, stands solved to an inferior position.

It deserves to be borne in mind that Inference and verbal testimony have been accorded the status of an independent tool of cognition. But recollection has been denied of this privileged position. It has not an iota or particle of probative value. It has right to claim equality of rank with any recognized cognitive tool. It is due to this obvious reason that Inference takes place despite the existence of recollection on the basis of its condition viz. dubiety.

Now we cannot forbear to consider another crucial question. It may be asked "does the existence of Inference or verbal testimony make the emergence of Inference impossible? The proponent answers –'yes'. The existence of the inferential judgement pre-supposes the presence of its conditions viz.

(a) dubiety of the probandum in the subject.

(b) observation of the probans (middle term) in the subject

(c) recollection of the invariable concomitance between the probans and the probandum.

(d) synthetic judgement (parāmarsha).

That is, the knowledge of the existence of the probans in the subject– and of the invariable concomitance between the probans and the probandum.

Now suppose, dubiety takes place in the first instance.

And conditions of inference referred to above occur in the succeeding instances. And the result confers that dubiety as the essential character of the subject would forfeit its right of existence. Since it is one of the cardinal principles of psychology that the succeeding qualities of the Soul entails the cessation of the proceeding one.

Dubiety and knowledge are regarded by the Naiyayikas as the specific attitude of the Soul. And consequently dubiety-signal must be liable to elimination by inferential condition which are the knowledge of the probans in the subject and of the invariable concomitance between the probans and the probandum. In this situation the very impossibility of the existence of dubiety qua condition of Inference rules out the possibility of inference. This argument is applicable to the verbal testimony with equal propriety. The consideration of the fact has persuaded the proponent to admit the possibility of the emergence of inferential judgement even in the face of the existence of recollection of the probandum in the subject. The cases of the existence of the inference or verbal testimony have been tacitly glided over due to the reason alluded to above.

The Neo logicians have subjected the view that dubiety is the essential character of the subject and as such a valid condition of Inference to a trenchant criticism. They have based their arguments on the unimpeachable evidence afforded by the verdict of psychology. A subtle analysis of the process of Inference involves a series of instances –

(a) 1st instance – emerges the dubiety of the existence of the probandum in the subject or knowledge-base.

(b) 2nd instance – is assigned to observation of the probans in the subject.

(c) 3rd instance – takes place the recollection of the invariable concomitance between the probans and the probandum.

(d) 4th instance – emerges the synthetic judgement (parāmarsha) as the immediate cause of inferential judgement.

Now the dubiety of the probandum which has occurred in the Ist instance becomes eliminated in the 3rd instance by the observation of the probans which has come into existence in the 2nd instance.

Here it has been remarked that the preceding specific attribute of the Soul ceases to exist with occurrence of the succeeding one. This detonates the thesis of the older logicians that – dubiety is the condition of Inference.

Since condition per se presupposes its precedence with Inference to its effect. Now if dubiety be regarded as the condition of Inference, it is bound to comply with inexorable Law of Causation as indicated above. But it has been shown that dubiety cannot exist immediately before its contemplated effect viz. Inference. And it is owing to this pulpable absurdity that dubiety cannot be accorded the status of pakshata qua condition of Inference.

The proponent puts forward a fresh plea to avert this conviction. He argues that the immediacy of precedence of dubiety is not essential for the emergence of Inference. Its mere precedence-be it mediate or immediate, is all that is regarded to make the materialization of inferential judgement possible. Here what is imperative is that – the dubiety as the condition of Inference must precede it in all circumstances. The question of immediacy of temporal sequel is obviously immaterial. But the plausible pleading would militate against vocal terms that Inference cannot take its rise in the event of the existence of the knowledge of the probandum in the subject or knowledge-base.

Further suppose that – dubiety occurs in a specific unit of time and the knowledge of the probandum follows it after the intermission of a long time. The precedence of dubiety is undeniable and as such complies with the condition laid down by the proponent. But is the proponent prepared to admit the possibility of emergence of inferential judgement in this particular occasion?

If he is faithful to his own position, his answer must be "No". So the theory – dubiety is the condition of Inference being the essential character of the subject – crumbles to pieces under the pressure of the criticism of the Neo-logicians.

The older logicians have partially admitted that the improbability of the immediate precedence of dubiety with respect to Inference detracts the value of their position. They have remoulded their genuine dispute to meet the present position. Further they have affirmed-although dubiety by itself cannot immediately precede the inferential judgement.

Yet such position would answer the purpose of facilitating its factors. So the facilitating factors of dubiety deserve to be regarded as the essential character of the subject and as such one of the conditions of Inference. An elucidation to this position is indispensable to reap its clear conception. The presence of the knowledge of the probandum and that of its contradiction in the intended subject, operates as an impediment to the emergence of dubiety. So their absence to be viewed as a factor which facilitates the cause of dubiety. To put it in other way round, the dubiety is conferred upon the capacity of coming into being by the absence of the 02 types of impediments alluded to above. The presence of the knowledge of the probandum and its knowledge is malefic or toxic to the occurrence of dubiety.

Inferential judgement cannot take place in the event of the presence of the 02 impediments (probandum and contradiction) mentioned above. The reason is too obvious to need elaboration. The existence of the 02 impediments implies the absence of the facilitating factors of dubiety i.e. the ostensible condition of inference.

Now even the amended definition of pakshatā is not immune from fault. The proponent has asserted that the absence of the knowledge of the probandum and that of its contradiction in the subject constitutes essential character of the subject. But it stands to clarification –

Is the one and identical absence of both impediments taken collectively in the 02 distinct cases of their absence considered distributively – intended to be predicated by the proponent? –

(1) 1st hypothesis – is not tenable. Since the presence of the one of the two impediments is compatible with the existence of the absence of both viewed in collective reference. And conformably to the case of the presence of the knowledge of the probandum and the absence of the both viz.

   (i) the knowledge of the probandum and

   (ii) the knowledge of contradiction – will be possible without the slightest mismatch. And this will amount existence of pakshatā and consequential materialization of inferential judgement.

   I cannot adjourn for a brief consideration of the logical concept of the absence of both (Ubhayābhāva) as adverted to above.

   An essential nature of both conjointly viewed involves its 02 component parts. The distinct identity of the 02 component parts becomes subservient to the whole. It is the felt fact that in the event of the presence of an earthen pitcher (Ghata), it cannot be asserted that the pitcher is absent with reference to the self-same spatiotemporal (spatial-temporal-input-cognition-output relationship) determination. But on the other hand, during the presence of the pitcher it cannot be affirmed that both the pitcher and/or the piece of textile/fabric are absent without entailing any contradiction. The consideration of this fact renders it apparent that the absence of the pitcher by itself is distinct from that of the both i.e. the pitcher and the piece of textile/fabric taken in their collection. This serves as the basis for contention that during the presence of the knowledge of the probandum – the knowledge of the absence of the both (probandum and contradiction) would be possible in the way indicated above.

(2) 2nd hypothesis – also does not stand on a better position. If 02 distinct cases of absence viz. (a) the absence of the knowledge of the probandum and (b) that of contradiction – are regarded as the essential character of the subject. This would be exposed to serious objection. Since the absence of contradiction is bound to be superfluous. Here the position need elaboration. –

The absence of contradiction has been asserted as one of the component factors of the essential character of the subject. And its implication is the inferential judgement which cannot take place in the event of the presence of contradiction. It is owing to this pulpable reason that the absence of contradiction has been held as one of the components of pakshatā which lies at the bottom of the contention that the absence of knowledge of the probandum is the essential character of the subject.

Now contradiction means – the knowledge of the absence of the probandum in the subject. And it is self-evident that this would act as an impediment to the emergence of inferential judgement. It has been authentically recognized that contradiction operates as a bar to the determinate judgement as a class. This fact amounts to or accounts for the essential nature of contradiction. Now it is undeniable that Inference is a determinate judgement. And consequently it is liable to be checked by contradiction in all circumstances. Besides contradiction is invested with an independent status of its own. And it is by virtue of this unique position that it deters the realization of inferential judgement. Further when contradiction foils inferential judgement by its inherent-cum-independent-right, the integration of its absence as one of the component elements of pakshatā is an obvious superfluity.

To be more precise the counteracting character of contradiction has been otherwise realized. Therefore, there is the slightest justification for the inclusion of its absence as linchpin of pakshata qua condition of Inference.

Let me look into the fresh definition by the few older logicians to evade the objections which are incidental to the view immediately adverted to above. They hold that the mere absence of knowledge of the probandum amounts to an essential character of the subject. This view is not the outcome of inordinate inclination of novelty. On the contrary it has been advocated by the exigency of the situation.

It has been shown in the previous paragraph that the absence of contradiction is not entitled to be regarded as an essential character or fitness of the subject. And as its consequence the absence of the knowledge of the probandum turns out to be the only claimant to the status of pakshata qua condition of Inference. But even this recast definition is not immune from problematic objection. The Upnishad has made the following proclamation:–

The Soul should be known by means of verbal testimony, Inference and contemplation. This significant utterance of the Upnishad has emphasized that the Soul which has been known through the medium of verbal testimony should be re-known through the affairs of inferential judgement. So it transpires on the ultimate analysis of the Upnishadic dictum that antecedent verbal testimony does not act as an contradiction to the culmination into inferential determination.

Were it not a possibility, the Upnishad would have been scarcely enjoined the inferable entity of the Soul in the face of the existence of its previous verbal knowledge. Therefore, the view that the absence of the knowledge of the probandum as an essential character of the subject qua condition of Inference would have foiled the categorical proclamation of Upnishad. Again there may be an object with respect to which exists a perceptual cognition. Here with reference to the self-same-perceived object, there occurs a desire for Inference.

This very fact will lead to the Inference of the object in question despite the presence of its prior perceptual cognition. Its truth is not liable to dispute.

We do have experience that desire is the most efficient condition in the matter of acquisition of knowledge. The existence of an object may be inferred on the basis of the probans. This execution or acquisition of inferential judgement of the object concerned may not answer the purpose of an inquisitive mind. And as such a sequence of it-an individual may resort to a fresh probans in order to attain the inferential knowledge of the self-same object. An instance will render the position clear.

Fire may be inferred in the hill by means of the probans (smoke). The process may be extended to one step further. The self-same fire may be inferred again by the exercise of the probans (light). And it is needless to observe that the subsequent inferential judgement takes place only in the event of the presence of desire or aspiration or fervor or urge or yearning or goal for its Inference with the help of the probans "light". This face-position also furnishes an ample proof that a predicate may be subsequently inferred notwithstanding that its previous Inference by means of a valid probans. So the original thesis of this group of logicians that the absence of the knowledge of the probandum (fire) is the essential character of the subject qua condition of Inference which stands assailed by unanswerable objections.

Another class of Naiyayikas has offered a novel definition of pakshata in order to evade the faults alleged above. It is asserted by the section of thinkers that the desire for inference is an essential character for the subject qua condition of Inference. This Neo-fangled Theory of pakshata will make possible the emergence of inference where it cannot take place due to the prior knowledge of the probandum in the subject as has been deserved earlier. A concrete instance may be cited for the clarification of the position. An inference of the Soul has been shown to be possible in the case of the existence of its verbal knowledge on the basis of the Upnishadic statement. In other way round, according to the terms of the recent definition of pakshatā – the inference of the Soul may take place as the matter of cause under the quest of desire for its Inference. It would sterilize the prior verbal knowledge of the Soul and will serve as an incentive to the emergence of the inferential judgement. We refrain from elaborating the Modus Operandi of the definition of pakshata under consideration with reference to other allegations to which its immediately preceding definition stands exposed.

The net result which has been achieved by the novel edification of the definition of pakshata may be expressed in a nut shell –

Inference has been shown to be impossible where there is the prior knowledge of the probandum. But according to the recent view – Inference will find no difficulty in its way of materialization if there be the desire of its Inference. Since it has been stated in the clearest term that the desire for Inference is pakshata qua condition of Inference. But a critical consideration reveals that the desire for Inference will not entitled to be treated as an essential character of the subject qua condition of Inference. The objection which explores the view that dubiety of the probandum is an essential character of the subject also equally overthrows the present dispute. An elucidation of the position is essential to bring home the allegation posed above.

Suppose for instance there occurs:-

(i) in the 1st moment – the desire for instances

(ii) in the 2nd moment – taking place the recollection of the invariable concomitance between the probans and the probandum.

(iii) in the 3rd consecutive moment – arises synthetic judgement in an identical way.

Now desire for Inference will cease to exist in the 3rd moment immediately after which takes place the effectuation of Inference.

The Law of Causality requires that the condition must immediately precede to its contemplated effect. But the desire for Inference as the ostensible condition of Inference is inept to comply with the Law of Causality which operates with arithmetical regard. The desire for Inference cannot be immediately antecedent to inference. The recollection of invariable concomitance which takes place in the 2nd moment entailing its cessation. Since it has been postulated by Naiyayikas that the immediately succeeding and perceivable attributes of the Soul bring about the auto-cancellation of the preceding ones. And it is undeniable that desire for both inference and recollection of invariable concomitance recites in the Soul as its perceivable qualities. This peculiar characteristic of the attributes of the Soul has already been adverted to in the case of the subject qua condition of Inference. It is remarkable that what has been said against dubiety as the condition of Inference is applicable to desire for Inference in equal degree.

But those who maintain the desire for Inference as an essential condition of Inference, are not prepared to surrender their position. It is attested by an undoubted experience that Inference takes place on the basis of the synthetic judgment persisting even after the cessation of desire for Inference in the manner as alluded to above. This argues out that a fresh desire for Inference has occurred immediately after the elimination of the previous one. This must be tacitly assumed. Since desire for Inference is an accredited condition of Inference. Under no circumstances it cannot be denied of this privileged status of its own. The very fact of the subsequent emergence of Inference indubitably reveals that the desire for inference is there. Granted that the desire for inference does exist in the situation under consideration.

But it is bound to presuppose the precedence of the knowledge that Inference is an infallible means to the realization of an end. And consequently this psychical phenomenon should manifest itself in an ordered sequence:-

(1) in the 1st instance – there is the synthetic judgement

(2) in the 2nd instance – occurs the knowledge of the inference as an unfailing means for the achievement of an end.

(3) in the 3rd instance – There arises the desire for Inference. Now it has already been remarked earlier that the synthetic judgement is the immediate cause of Inference. But in the present occasion it cannot immediately precede the contemplated Inference. Because it will cease to exist in the 3rd instance which has been assigned to the desire for Inference. The reason for the cessation of synthetic judgement has been elaborated in the preceding paragraph. So even the admission of the fresh desire for Inference will be of no avail due to the absence of the synthetic judgement. But this objection is based on a misconception of the whole situation.

It is the resultant inferential judgement which is responsible for the pre-supposition of a novel desire for Inference. And in an identical manner, a fresh synthetic judgment, also, will be involved by it after the emergence of the fresh desire for Inference. It is the resultant Inference that chiefly leads to their presupposition (Purvadharana) in the sequences indicated above.

To be more precise – an Inference is there. It is bound to remain unaccounted for without the assumption for fresh desire for Inference and synthetic judgement. To sum up, it is the desire for Inference that should be regarded as an essential character of the subject qua condition of Inference. And the credibility of this view cannot be challenged for the reason alluded to above.

But the argument pleaded by the opponent only betrays his levity of thought. It is not frequently felt that the peal of thunder leads to the Inference of cloud. This inference is spontaneous in its essential nature and is characterised by the accredited condition of Inference, namely:-

(i) observation of the probans

(ii) recollection of universal concomitance

(iii) the synthetic judgement

Now Inference takes place despite the absence of its desire.

A concrete instance will make the position clear-Suppose an eye-catching procession burst upon the view all on a sudden. It invariably causes the Inference of a festive occasion. It is needless to observe that in the present case – the desire for inference is obviously absent. These considerations make mincemeat of the contention that desire for Inference is an essential character of the subject qua condition of Inference.

Earlierepisodepredatesafaithfulexpositionof NN Gangeshopadhyaya's criticism of the view of the older logician on the essential character of the logical subject.

In the present chapter an attempt will be made to lay bare his subtle and sublime speculation on the problem under probe:-

Let me define an essential character of the logical subject framed by the Neo-logician Gangeshopadhyaya as follows:-

The essential character of the logical subject lies in the absence of the conviction of the probandum which is accompanied by the absence of a desire for Inference. What is intended to be conveyed may be put forth in the following orders:-

(1) Inference will take place in the event of the presence of both – the conviction of the probandum and the desire for Inference.

Truly, the conviction of the probandum-a deterrent factor of Inference is there. But it confirms to the terms of the definition in hand. Because the conviction of the probandum is not attended with the absence of the desire for Inference. It deserves to be stressed that the conviction of the probandum unaccompanied by the absence of a desire of Inference cannot operate as a bar to Inference. To put it in otherway round, the conviction of the probandum accompanied by the absence of desire for Inference counteracts its materialization. Now in the case under review– the presence of both – the conviction of the probandum and the desire for Inference renders it evident that the conviction of the probandum as determined by the absence of the desire for inference is not there. That is to say, the absence of the adjective viz. the absence of desire for Inference culminates into the absence of the determined conviction of the probandum as such. And such situation fulfills the condition of pakshata as laid down in the definition.

(2) Inference is permissible if there be the desire for Inference and absence of the conviction of probandum. The presence of desire for Inference involves the absence of the adjective -viz- the absence of desire for Inference. To be more explicit, both the adjective and the substantive (conviction of the probandum) are absent in the present situation. So it conforms with the phraseology of the definition.

(3) Inference is allowed when there is the desire for Inference. In the present case the adjective viz. the absence of the desire for Inference is there. But the substantive – conviction of the probandum is obviously absent. The ultimate upshot is that the substantive which occupies a superior position is absent. And consequently it entails the absence of the determined one despite the presence of the adjective (the absence of desire for inference) referred to above. Therefore, it serves as a perfect argument with the condition laid down in the definition.

Now a resume is felt essential for the expression- "the absence of desire for Inference" – occupies the position of an adjective. It has a distinct utility of a great moment. The conviction of the probandum deters the realization of Inference. But in the event of the presence of desire for inference it ceases to be deterrent factor. It is owing to this reason that the mere conviction of the probandum has not been regarded as a bar to Inference. On the contrary, when the conviction of the probandum is accompanied by the absence of the desire for Inference it acts as a prevention. That is to say, its determined conviction becomes endowed with obstructive-efficiency. This consideration has caused to insert the proviso – the absence of the desire for Inference in the body of the definition. It is the prospect of the achievement of this objective that has led to the introduction of adjectival expression – the absence of the desire for Inference.

But this expectation cannot be fulfilled in the way adverted to above.

Since the desire for Inference has been accorded, so to speak, the position as a stimulant (uttejak) to make the materialization of Inference possible even during the presence of conviction of the probandum indispensable before the actual drawing of Inference. Since its absence will be equivalent to that of the deterrent element. Consequently, Inference will legitimately be expected to take its rise on the basis of its accredited condition viz. synthetic judgement. A priori consideration makes it imperative that the conviction of the probandum, the desire for Inference and the synthetic judgement must be co-present in the identical moment. Its non-compliance will divert the adjective viz. the absence of desire for inference of its intended utility. But a critical examination of the whole situation discloses that there is not even the remotest prospect of its success. It is an admitted fact that the psychical phenomenon as a rule cannot last for more than 02 consecutive moments:—

(1) in first moment - it is brought into existence

(2) in second moment – (the immediately succeeding) is known of the moment of its existence.

(3) in third moment, it passes out of its existence.

It is, however, permissible that a new psychical phenomenon may appear in the moment of existence of a preceding one. The net result is that the co-existence of the 02 psychical phenomena is possible in the self-same moment. But under no circumstances the rigor of psychical law will permit the co-presence of a set of 03 phenomena in an identical moment. The very impossibility of their synchronous occurrence renders the adjective – the absence of the desire for inference-futile. Consequently, there is not the slightest warrant for investing the desire for Inference with the position of the stimulant.

Paraphraser of Neo-Logic Raghunath Shiromani in his rejoinder has observed that objection is based on an ignorance of the logical situation. He has mentioned 02 specific occasions as:-

(a) the justification of the adjective

(b) the absence of the desire for Inference

In the 1st moment-there occurs the desire for Inference.

In the 2nd moment-arises a distinct type of recollection –the hill which holds fire, is holding smoke – (the sub-concomitant of fire). Evidently the judgement in the form of the hill holding smoke (the sub-concomitant of fire) is called synthetic judgement. And the determination – "the hill holds fire" is technically termed as the conviction of the probandum. The integral parts have been resolved to exhibit the precise nature of the conviction of the probandum and synthetic judgement. Both of them are produced for their distinct conditions and as such are invested with an auto-status of their own.

But in the case under review, their accidental association has occasioned a one and identical type of cognition. Their distinctive individuality (self identity) or character or self-hood, so to speak, has been completely merged in the self-same cognition.

Viewed in its 02 distinct aspects, this cognition is, also, entitled to be called either a prior conviction or a synthetic judgement. Its ultimate upshot lies in the 2nd moment which is the moment of existence of the desire for Inference. Both conviction of the probandum and synthetic judgement will be present as a matter of course. The 02 distinct moments are assigned to conviction of the probandum and synthetic judgement. Now the integration has made their allotment to the self-same moment possible. In other words the utility of the 3rd moment has been totally dispensed with.

The 2nd moment becomes the pivot around which the entire process of Inference revolves. The invention of the preculiar situation was dictated by a logical compulsion. Since the alleged charge of the impossibility of Inference due to the impossibility (that) of synchronous or coeval presence (Sahkalin) of 03 accredited conditions stands completely countered or retaliated. So the adjective – "the inference of desire for inference" has been introduced to pave the way for actuation of Inference in the situation adverted to above. As a cause of which the present occasion – the desire for Inference and the distinct type of recollection are present in the 2nd moment. But the adjective – the absence of the desire for inference is evidently absent. And it is one of the fundamental postulates of logic that the absence of the adjective entails that (absence) of the determined substantive. So the absence of the former terminates into the absence of the latter. And such absence constitutes the logical character of the subject qua condition of Inference. Now as the presence of this adjective which has been inserted into the fabrication of the definition facilitates the drawing or layout of Inference and its essentiality is not liable to counter.

Further to elaborate, in the 1st moment-there occurs a peculiar type of recollection which embodies in itself as:-

(a) synthetic judgement
(b) the conviction of the probandum
(c) the knowledge that inference is a means to the realization of an end.

In other words – this recollection should be viewed as a tout ensemble of distinct elements – the synthetic judgement and the rest. It has already been observed that a peculiar type of knowledge takes place in the event of the collocation or juxtaposition of the accredited conditions of synthetic judgement and conviction of the probandum with the knowledge that the inference as a means to the realization of an end. The nature of this recollection may be correctly presented in the following logical mode. –

The hill which holds fire is also holding smoke-being the sub-concomitant of fire, and the Inference of fire is conducive to the access to my intent.

Now in the succeeding moment – there arises the desire for Inference which ultimately terminates into the consequential drawing of Inference. It is true that the thwarting agent of Inference namely – the conviction of the probandum in the shape of recollection is there. Yet the desire for Inference as a stimulant paves the way to the materialization of Inference.

The function of the stimulant renders a deterrent factor of Inference to be idle and inoperative. Apart from that, if facilitates the smooth passage of cognitive operation involved in Inference. It should not be the lost sight of that synthetic judgement in the accredited condition of Inference. But in the present case the recollection has been accorded the function of synthetic judgement. Now the consideration of the whole situation warrants the computation of all the essential conditions of Inference. It is but legitmate to express its occurence as a matter of course. Now if the adjective – "the absence of the desire for inference" be deleted from the definition, this would mean the absence of pakshata – the essential condition of Inference. And this would deter the emergence of universally recognized Inference in the case under discussion.

It is manifest that

(i) the omission of the objective and

(ii) the absence of the desire for Inference would bring about an undesirable consequence in the present situation. It would impede the kit of inference in the case alluded to above. And it is needless to mention that expectation of the occurrence of Inference in this situation is a legitimate one.

The question may be exposed – "why the exclusion of the adjective will make the contemplated Inference impossible. The answer is crystal clear – the unique type of recollection referred to above, is of the nature of the conviction of the probandum. And consequently it is deterrent to Inference. So as an immediate effect or upshot, its absence can facilitate the path of Inference. And this would ultimately constrain the admission that the absence of the prior conviction of the probandum being the condition of inference. Now in the case under discussion-recollection which is identical to the prior condition of the probandum will thwart the emergence of the intended Inference. And it is with a view to avoiding undesirable contingency or occurrence or eventuality that the adjective in question has been inserted into the definition. Its collaboration or inclusion or integration will be conducive to the fruition of the essential character of logical subject.

Because the desire for inference is there. And it is tantamount to the assertion that there is not the absence of desire for Inference. And the absence of the desire for Inference is tantamount to the absence of the adjective. And it has been pre-observed that the absence of the adjective terminates into the absence of the determined-substantive accompanied by the absence of desire for Inference. It is now evident that the synthetic judgement per se is the immediate condition of Inference. And it is also, equally apparent that this synthetic judgement has been assumed to be identical with the conviction of the probandum in order to demonstrate-how does the desire for Inference function as a stimulant. The very nature of the stimulant involves the presence of a prevention. A stimulant will cease to be stimulant if there is no deterrence. But the question – why this synthetic judgement has been regarded as of the nature of the knowledge of the means to the realization of an end – is to be discussed. A brief resume is essential for a clear cut conception of the situation. It has been shown in the 1st situation that the desire for Inference takes place in the 1st moment and the distinct type of recollection arises in the 2nd moment. This sequence has been completely revised in the 2nd logical situation. It is remarkable that in the 2nd predicament – the desire for inference takes its rise in the 2nd moment. There is a deeper reason to account for the reversal of order which has been adopted in the 2nd situation.

The desire for Inference which has been stated in the 1st situation as occurring in the 1st moment makes its imperative that the knowledge of means to the realization of an end must be treated as its antecedent. But the latter is an effect of the former.

The Law of Causality requires that the cause and antecedent invariably precedes effect or the consequent.

So the assumption of the previous existence of the knowledge of the means to the realization of an end turned out to be unavoidable logical compulsion. This assumption of an auto-status of the knowledge of means to the realization of an end entailed redundancy of superfluity of thought.

In order to avert the flaw, the 2nd situation has been reversed with essential amendment or modification. The synthetic judgement and the knowledge of the means for the realization of an end have been merged together to achieve a numerical unity between them. It is owing to this merger, the synthetic judgement has been regarded as the nature of the knowledge of means to the realization of an end.

In short, it partakes the mode of expression that Inference is the means to the realization of an end. This unity has affected a simplicity in comparison with the preceding situation. It has an additional advantage to its credit. The appearance of desire for Inference in the 2nd moment has conferred an element of the novelty upon it. This 2nd instance illustrates that the desire for Inference may function as a stimulant whether it precedes or succeeds the conviction of the probandum.

The 1st instance was conceived on the basis of the precedence of the desire for inference. But it did not include the accredited case of inference which occurs even when it succeeds the conviction of the probandum. Certainly it would have betrayed lacking of logical acumen. It is owing to this reason that these 02 situations of different description have been put forth by the logicians –

But a relevant poser arises.

The preceding 02 situations are constituted of recollection. In other words, recollection has entered into their very composition or structure. What is the reason for preferential treatment towards recollection? The answer calls for a pro-tem digression from the main subject of investigation.

The condition of Inference is synthetic judgement. The avowed acceptation of this expression is a specific type of knowledge – a knowledge as such admits of two-fold classifications:-

(i) experience

(ii) recollection

Now if recollection were superseded by experience, it would have entailed complexity of thought. Since the experience is judgemental in character, the judgement (Vishishta Gyan) in its term is susceptible of being resolved into 02 component parts. (i) Adjective (ii) Substantive.

In other words, the substantive-adjectival type of relation is that which constitutes the judgement in all cases. It is the fundamental rule of thought that the knowledge of the adjective is essential for the arousal of the judgment in question.

Suppose – experience is substituted for recollection. Then crux of intrinsic or non-intrinsic quality of recollection turns on that of experience. And such experience is acknowledge through 07 causes:-

(1) padgyān (Textual Knowledge)

(2) Padpadharth Sambandha Gyan (Relation of location and meaning of things enumerating –

16 - Naiyāyikas

07 – Vaisheshikas

25 – Sankhyas

27 – Patanjali

(3) Padārtha Smaran Ākānkshāgyan – aspiration for recollection of things.

(4) Āshakti or Anāshakti (fondness or nonfondness)

(5) Yogyatā Gyan – Compatibility of Sense knowledge

(6) Tātparya – Purport

(7) Tatparyagyāpak prakaran – purport of precepting topic.

So to say recollection is different from concepts and confusions being extra inquiries. The recollection depends on acquisition, consolidation and retrieval. Whereas experience depends on manifestation, integration and moving information. Further experience ensures perfection through direct observation or participation in events. Collectively the experience encompasses the involvement of the knowledge of the adjective as a pre-condition of its occurrence. And it is needless to refer to the fact that such situation would bring down plethora of thought. But recollection is not a judgement. Therefore, the consideration of simplicity has caused its retention in the fabrication of the 1st and the 2nd logical situation.

The 02 preceding logical situations are based upon the assumption that the synthetic judgement is of the nature of recollection. But it cannot be gainsaid that synthetic judgement may also partake the nature of experience. Consequently its omission for the 02 aforementioned situations is felt as a serious drawback.

So to test the necessity of the application or exercise or service of the adjective causing concomitant with the absence of the desire for Inference (3rd situation) has been formulated on the basis of the presumption that synthetic judgement is of the nature of experience. This case may be recorded in the following order:-

Suppose the recollection of the adjective occurs in the 1st moment. The desire for Inference and the contact of the sense-organs with the logical-base occurring in the 2nd moment. The observation- the hill holds fire" – the probandum and the probans as its concomitant takes place in the 3rd moment. Now an analysis of the present situation reveals that there are desire for inference and the conviction of the probandum. Now the adjectival clause coincided with the absence of desire for Inference has been left out from the definition of pakshatā.

The universally acknowledged Inference will be debarred from occurring due to the presence of the conviction of the probandum. But the integration of the adjectival clause will lead to the drawing of Inference without countering any argument. Because in the situation under hand – there is the desire for Inference. And consequently the conviction of the probandum due to the absence of the desire for inference ceases to be an impediment to inference.

I have woven 03 webs of logical situation. So far the 02 preceding cases are concerned with the connective tissues are self-transparent. But the 3rd situation appears to all intent as a dark maze. So its unfoldment is felt essential for clear understanding – it is evident that the recollection of the adjective has been conforming a position which is antecedent to the desire for Inference. The assignment of their respective positions is dictated by a logical compulsion. Since there are only 03 possible alternatives which may be resorted to the allocation of their position enumerated as follows:-

(1) Both of them maybe co-eval with respect to their occurrence

(2) The recollection of the adjective may yield precedence to the desire for inference

(3) Lastly the recollection of the adjective itself may precede the desire for Inference.

Now the 1st hypothesis may be easily disposed off. In as much as it is one of the cardinal canons of Logic that there must have a co-existence of knowledge and desire. The 2nd hypothesis-also cannot bear scrutiny. The precedence of the desire for inference involves 03 consecutive psychical operations-

The 1st moment-there occurs the desire for inference

The 2nd moment-arises the recollection of the adjective in question

The 3rd moment-takes place the synthetic judgment qua conviction of the probandum.

Now the desire for Inference will cease to exist in the 3rd moment which has been assigned to the synthetic judgement. The cessation of the desire for Inference is bound to entail the resultant impossibility of the character of the logical subject which renders the adjectival clause-

"the absence for the desire for Inference" – irrelevant. It should be borne in mind that - 3 logical situations have been invented to demonstrate the result of the insertion of the adjectival clause repeatedly adverted to above.

Now the 3rd alternative is adopted to meet the requirement of the situation. And this fact has been clearly demonstrated in the preceding paragraph.

There is another point of equally great moment-the desire for inference and the contact of the sense-organs with logical subjects have been held to the occurring in an identical moment. There is the deeper reason for maintaining their co-existence. If the contact of the sense-organ with logical subject were accorded priority over the desire for inference undesirable consequences would appear in its wake. The alleged contact is an accredited condition of ordinary perceptual cognition. Its presence would counteract the desire for inference and prepare the way for the emergence of perceptual cognition.

It is one of the fundamental principles of Logic that the assemblage of the totality of the conditions of ordinary external perception is invested with superior power. We had occasion to observe that such ordinary perceptual cognition would be the nature of the conviction of the probandum in the logical subject. Not only this, there is also the obvious absence of the desire for Inference. This would render the emergence of inference as downright impossibility. This would ultimately force the admission of the superfluity of the adjectival clause in question. These considerations are responsible for denying pro-tem antecedence to the contact of the sense-organ with the logical subject.

Again a pertinent poser arises- the desire for inference has been accorded the status of stimulant. That is to say, the prior conviction of the probandum becomes deprived of its anti-function during the presence of the desire for Inference. In one word, it provides a powerful incentive towards the drawing of Inference in the teeth of deterrence which appears in the shape of an antecedental awareness of the probandum to the extent of posing complications.

In the 1st instance – arises the synthetic judgement which of the nature of prior conviction of the probandum.

In the 2nd instance – the succeeding one occurs the desire by Inference which expresses in itself by the mode of expression. Let there be an awareness with reference to something.

In the 3rd instance – admittedly the situation conforms to the view alluded to above making the emergence of Inference inevitable. Since the desire for awareness of something includes within its fold the desire for Inference. The desire for awareness as a class is wide enough to include the desire for specific type of knowledge – Inference. But inference cannot be permissible in the case under review.

Neoparaphraser Gadādhar Bhattacharya has advocated to account for the non-emergence of Inference in the present instance. He observes that desire for Inference per se cannot provide stimulus or fillip to the drawing of Inference in all cases alike. It is attested by the evidence of psychology that desire becomes specified with achievement of the desired effect. Consequently desire offers incitement or stimulation to that which is instrumental to the realization of an end. It is due to this patent fact that a deterrence is rendered inoperative by its mere presence. In the situation in hand the prior cognition of the probandum being desire for knowledge of something is there. This desire cannot act as a stimulant with respect to the Inference of fire in the hill. Because a consideration of the nature of desire exhibits that it is not interested with any particular type of knowledge.

In other words, any specific variety such as perceptual or inferential is adequate to bring about the fruition of the desire in question. So it cannot claim to be stimulant with reference to the Inference of fire. If it be so, why does it fail to act as stimulus upon Inference when its stimulating capacity has not been totally denied even concerning inference? Here the answer is transparent – the conditions which are given in the case under consideration are competent to engender their mental cognition. And this will operate as a complete satisfaction of the desire for knowledge of something as alluded to above.

Now when such desire has been set at rest in the manner indicated above. There can be no justification for resorting to an inference of the previously known object to its satisfaction.

A serious problem pokes up by itself for solution – the immediately preceding logical situation has been recorded in the following order:-

(i) An antecedent cognition of the probandum

(ii) A desire for knowledge of something

But such set-up is the outcome of naive fancy. Since the Law bears sway over psychical phenomena testifying to the fact that a desire ceases to be realizable of the desired object.

To put it in other way round, an antecedent realization of an object acts as a bar to the emergence of its desire. In the present instance the desire is for the knowledge of something. But the knowledge of the probandum is already there as a settled fact. And it is entitled to be considered as an object of desire which has for its object – the knowledge of something. In as much as this knowledge of something grasps the antecedent cognition of the probandum to which has been assigned the 1st and the foremost position.

In other words, the desire for knowledge of something stands fulfilled by the antecedent conviction of the probandum. Therefore, with the satisfaction of the desire there is not the faintest possibility of the desire for knowledge of something. To express it in a strict logical form, the satisfaction of desire by means of the prior awareness of the probandum operates as an impediment to the subsequent desire for knowledge of something. So the objection raised on the basis of such situation is devoid of substance.

But Neo-logician Gadadhar Bhattacharya is aware of this flaw.

So he has offered an amended mode of explanation which may be set forth as follows.

Let there be the knowledge of something immediately succeeding to this viz- prior conviction of the probandum. When the desire is concerned with the subsequent knowledge of something the preceding awareness of the probandum cannot act as deterrent. Since the preceding knowledge of the probandum is a fait accompli. In a critical tone antecedently apprehension of knowledge cannot be an object of successive desire. And so it is evident that the principle – the preceding realization of an object operates as a bar to the emergence of a desire for the same – is not applicable to the situation adverted to above.

Now it is manifest that the desire for Inference by itself cannot be effective of the intended result. Since it stands assailed by serious objection alleged above. This has impelled oriental logician Raghunath Shiromani to express it by a remodelled form of expression as follows:-

The desire for inference implies the desire having inference as its object and referring to its (Inference's) adjective in the form of the fact of having the intended logical subject which stands qualified by the intended predicate.

The mentees of Neo-Logic are earnestly required to preserve the patience of his mind in order to follow the wise or deep implication of this condensed statement. A concrete instance is expected to render the position intelligible. Suppose, the desire assumes the following form – "Let there be an Inference of fire in the hill". The desire expressed by means of this averment evidently bears out the truth that the desire plays the role of stimulant. Since the object of this desire is evidently an Inference. And the Inference in its turn is conversant of the intended logical subject-hill qualified by an inferable predicate (fire). And the fact that the qualified subject being the object of Inference is taken stock of by the desire. This conforms to the condition required to make the desire stimulant or excitant. This interpretation of desire for Inference evades the objection framed with Inference to the desire – let there be an awareness of something. Because neither the logical subject nor the predicate appears as an object of this desire.

But Neo-logician or paraphraser Jagadish Tarkalankar, however, has put forth his own independent explanation. He observes that if the pure and simple desire for inference be regarded as a catalyst, this would occasion an absurd consequence.

Suppose, there is a synthetic judgement qua conviction of the probandum. And immediately after it, there occurs the desire – let there be an awareness of the substance-hood. In the presence instance the Inference of fire is bound to claim smooth passage to the compliance with all the essential conditions. Such awareness as genus (Jati-class) includes the inferential awareness as its species. Consequently, it conforms to the condition essential to be accounted as a desire for inference properly. In this eventuality the Inference of fire is bound to occur without any let's or hindrance. Certainly in this situation 'fire' ought not be inferred for reasons stated below:-

The previous cognition of the 'fire' is present as a palpable prevention. Furthermore the 'desire' in the present context be expected to play the role of catalyst. Because it has not the Inference of fire as an adjective datum. This desire cannot be satisfied by the awareness of substance accompanied with substancehood. Even the occurrence of the apprehension of fire together with firehood cannot fulfil the desire in question. Under the circumstances, the Inference of fire cannot be permissible on the basis of this desire assuming the role of stimulus.

Now it is evident that the 02 paraphrasers have adopted distinct attitudes in the matter of expounding the self-same expression of Raghunath Shiromani and Jagadish Tarkālankar. Here Jagadish Tarkalankar has located a serious flaw in the exposition of his predecessor and as such made a clue for correction.

The whole problem hinges upon the following vital question – is the desire for inferential awareness of something possible immediately after a synthetic judgement qua conviction of the probandum?

It is attested by the testimony of psychology that the existence of preceding awareness holds the subsequent desire for its emergence in abeyance.

It stands to reason that with the realization of an adjective, the possibility of occurrence of the fresh desire pertaining to it becomes completely thwarted. It has been reiterated over and again that the former operates as a deterrent to the latter. If this be the real state of affairs, how can the observation of Raghunath Shiromani be permitted to stand?

The 3rd Neo-exponent Gadadhar Bhattacharya has attempted to reconcile the incongruity by employing the adjectival phrase – "immediately succeeding before the intended awareness". The result derived by this device has been dwelt upon in the preceding paragraph.

Neo-critic Jagadish Tarkālankar confers his own solution to this problem in hand. And it has already been represented in a clear-cut way.

I cannot withhold or abstain from expressing my own view on the problem under discussion. I have abundantly made it clear that Noe-Critic Gadadhar Bhattacharya maintains that there is no chance of the occurrence of desire for the awareness of something – if it is immediately preceded by the cognition of the probandum. The initiation of the phrase – "immediately succeeding" in order to circumscribe the awareness fails to solve the problem. Because any cognition, intended after it, is bound to be succeeding one even without the application of the adjectival phrase posed by Gadadhar Bhattacharya.

It is apparent that this adjective is assuredly superfluous one. So the arbitrary ascription of successiveness to a future awareness cannot make it a desired one.

The problem is too recondite to be given a wide berth. So I feel called upon to deliver it deeper in the interest of thoroughness and precision. It may be argued in the present context that a previous cognition cannot act as a bar to the desire for a succeeding awareness. Since there can be no conflict between before and after in the situation under consideration. The priority and posteriority in the matter of succession avert the prospect of any class between the 02 types of cognition.

The desire in question comprises in its fold viz. cognition and its posteriority. Consequently, it steers clear of the jurisdiction of the pervious conviction and thereby insulting from the sphere of its objective datum. The "would be desire" falls beyond the range of operation of the thwarting agent – the previous conviction.

So it appears to all intents intending that the plausible plea places the position of the Neo-Critic Gadadhar Bhattacharya on a secure foundation.

I have critically reflected on the defence laid down in the preceding paragraph.

And I venture to aver that it is not absolutely immune from hostile criticism. Now Neo-Analyist Gadadhar Bhattacharya – has explained in extenso the nature of the desire by prefixing the adjective successiveness to the intended awareness. This desire is bound to be satisfied by the occurrence of awareness of the object which has been known immediately before. So the attempt at bringing about an element of novelty by the insertion of the adjective which was urged against the interpretation of Gadadhar remains unanswered. The position calls for amplification –

It is an avowed psychological phenomenon that the desire for awareness occurring after a preceding one (awareness) can be fulfilled either by an awareness specifically other than the antecedent one or by an awareness referring to a novel objective datum. That is to say, the element of novelty must pertain either to the species of awareness or to the objective datum with which it is conversant. It should not be the lost light of which the desire has for its content-both the awareness and its objective datum. The failure to comply with either of these 02 alternatives makes the desire liable to be checked by the antecedent awareness.

An instance would drive the nail –

Suppose there takes place a synthetic judgment –the hill holds fire and the smoke as its subconcomitant (Vyapya). Assuredly this judgement is of the nature of the antecedent conviction of the probandum. In this context, there does emerge the desire for emergence of fire. Ponder over the fact that so far the objective datum is concerned, the element of novelty is conspicuous by its absence. In as much as the existence of fire in the hill has already been known by the synthetic judgement qua previous conviction of the probandum. But Inference is a distinct variety of cognition from that alleged cognition of the probandum. In an identical manner, otherness of the object, also, serves as an incentive for the desire for awareness.

It is evident by experience that the desire for the cognition of a piece of the fabric or textile does not take place after the cognition of a pitcher. There is neither the specific variation of awareness nor the difference of its objective datum in the case in hand. This consideration renders it apparent that an awareness dissociated from reference to an objective datum cannot take place subsequent to a previous conviction.

Again a problem crops up – it has been stressed that post-awareness-desire for awareness may occur on the compliance with the 02 conditions indicated above. But it is felt that charming beauty occasions desire for its consecutive and repeated cognition. Admittedly there is neither the distinctness of objective nor the specific variation of cognition. So the principle adverted to above is devoid of foundation.

Considered in the light of the logical device of Neo-critic Gadadhar Bhattacharya appears to be correct one. But this argument is not tenable. It should be borne in mind that awareness of a means as conducive to an end is the pre-condition for the emergence of desire. The repeated observation of a thing of beauty is felt instrumental to the realization of an end. It brings delight in the mind of the observers. In the case under review, awareness refers chiefly to the beautiful trait or feature of the object. So the reference to a distinct object is not liable to dispute. The attraction of the beautiful thing is responsible for the emergence of desire despite the absence of the alleged 02 conditions. The spell of beauty has been unanimously recognized by the thinkers of different schools of Indian philosophy. This is a mere exception and as such cannot be resorted to for decrying the principle exposed above.

I am not holding a brief in favour of anyone of these 02 exponents of Neo-Logic. A critical reflection has convinced me that the interpretation presented by Jagadish Tarkālankar deserves a preferential treatment. Since he has clearly indicated the content with respect to which the future desire is to take its rise. This has eliminated the prospect of the impossibility of the emergence of desire and as such has this advantage to its credit. But the logical device put upon by Neo-Logic critic Gadadhar Bhattacharya turns out deterrent to the very possibility of desire.

In the preceding paragraph, the discussion centres around the respective position of previous conviction and the desire for Inference. The former precedes the latter, whereas the latter succeeds the former. The consideration of this question of the precedence and succession of them has been primarily responsible to take in the interpretation of Jagadish Tarkālankar in a favourable light.

But a reversal of their order of succession may put the argument of Gadadhar Bhattacharya in an advantageous position-

Suppose, the desire for awareness of something occurs in the 1st moment. And in the succeeding one there takes place the synthetic judgement qua conviction of the probandum. Now in this situation, the presence of the previous conviction qua impediment is undeniable. But owing to this existence of the excitant in the shape desire – Inference will be irresistible. To be more explicit, the character of the logical subject qua condition of Inference has been defined as the absence of the conviction of the probandum in the logical subject-accompanied with the absence of desire for Inference.

The present case conforms to the phraseology of this enunciation of the condition of inference qua character of the logical subject. The true presence of the conviction of the probandum is not liable to dispute. But there is the absence of conviction of the probandum accompanied with the absence of the desire for inference. Certainly the drawing of Inference in the present occasion has not received the seal of approval of the logicians. Since the desire for awareness of something is followed by the previous conviction of the probandum. So this desire stands satisfied by the immediately succeeding conviction of the probandum. Consequently, the deduction of inference has been bereft of its justification.

Now it would not be unfit to observe that the expression – the desire for inference has been inserted in the definition of the credential of the logical subject.

Further Neo-Logic critic Gadadhar Bhattacharya being the exponent of Tattvachintamani "Deedhiti" has observed that there implies the desire for Inference of the intended predicate in intended subject. This has been elaborately treated off in the course of dissertation. Besides he has remarked ad rem that inspite of the presence of the conviction of the probandum, the undesirable entailment of inference will be inevitable on the basis of the desire for awareness of something in the event of non-acceptance of the qualified desire for inference referred to above.

The exposition of Neo-logic critic Raghunath Shiromani is perfectly sensible. So what impelled Jagadish Tarkālankār to innovate the situation as a vindication of the aptitude to the desire for inference.

Neo-Logic- crtic Jagadish Tarkālankār, himself has furnished a feedback to this query. He has anticipated ipse dixit of Raghunath Shiromani in order to reinforce or prop us this position-that desire deserve as a catalyst to the object of which is not realized after the occurrence of it (desire). Viewed in the light of this observation of Neo-Logician Raghunath Shiromani, the newly devised situation cannot be a warrant for the adjective indicated above. Since the priority of occurrence has been accorded to the desire and posteriority to that of the conviction of the probandum. This conviction is the object of desire for awareness of something. Therefore, the desire cannot be expected to play the role of the stimulant. Consequently, the conviction of the probandum will act as a prevention. So Jagadish Tarkālankār's modification of mode of expression of the desire has resulted in the creation of a novel situation.

Further Neo-Logic critic Raghunath Shiromani has laid bare the essential criterion of the desire which serves as a stimulant – his cryptic representation is pregnant with momentous consequence. It was needed to meet the complex logical situation which has been exposed below-

Ponder over the unique type of recollection which embodies by itself synthetic judgement and awareness being is instrumental to the realization of an end. Its double character may be demonstrated by the following mode of expression – The logical subject holds the probandum, its sub-concomitant, the sub-concomitant, of its sub-concomitant, and the inference of the probandum and that of its sub-concomitant are instrumental to the realization of an end. This recollection in question is a singular awareness referring to several objects at one time. It is obviously tout ensemble (Samuhavalamban). It presents its unity and is competent to exercise the function of several awareness.

Now it transpires on scrutiny that recollection alluded to above is one of the natures of synthetic judgement qua condition of Inference of the probandum and its sub-concomitant. In as much as this recollection due to making Inference to the sub-concomitant of the probandum in the logical subject occupies the position of the synthetic judgement. And owing to its making reference to the sub-concomitant of the sub-concomitant of the probandum, it stands identified with the synthetic judgement which is the accredited condition of Inference of the sub-concomitant of the probandum. So to the essential disposition of the 02, the synthetic judgement exists in the recollection.

Now it is in order to meet the counter intuitive or farcical fallout that Neo-critic Raghunath Shiromani has prefixed the adjective – "unqualified of the conviction of its own object" to the desire intended to act as a stimulant – considered in the aforementioned situation which stands completely precluded.

Since the desire in the case, under dispute, has been deprived of its stimulative virtue due to the occurrence *of the 1st Inference*. Besides it is remarkable that the 1st inference has already been asserted to be of the nature of previous conviction of the probandum. So both the absence of the stimulant and the presence of the previous conviction will hold the contemplated *2nd Inference* in abeyance. Therefore, the need of the insertion of the adjectival clause before the desire for inference stands completely justified. Now the degression lays bare that the situation devised or innovated by Neo-critic Jagadish Tarkālankār as a justification for criterion predicted by desire is founded upon a rational ground. Again a fresh dubiety crops up – the nature of the desire qua stimulant as elucidated above is open to serious lacuna –

Ponder over the desire – let there be perceptual cognition of the hill holding fire. It is a case of error. Since an Inference has been erringly identified with perceptual cognition. And this furnishes the ground for a unique type of desire.

Here it should not be overlooked that a genuine Inference has been mistaken for perceptual cognition. The consideration of this fact entitles the present desire to be regarded as one-having Inference as its adjective datum. So there would be the undesirable entailment of inference on the basis of this desire despite the presence of the previous conviction of the probandum. It is the felt fact that the distinction of the object is referred to by desire. Besides the cognition which takes its rise on the basis of this desire is expected to own this very character referred to by desire which cannot be desirable.

To be more accurate, the character "perceptiveness" falls under the consideration of desire. Consequently it is legitimate to anticipate its presence in inference. But the expectation stands belied or contradicted.

Here Inference taken by itself is a diametrical opposite or antithetical to perceptual cognition. So the perceptiveness cannot reside in it. Therefore, one is driven between the horns of dilemma. Both commission and/or non-commission of the criterion intended to confer the status of a catalyst upon desire are redded by difficulty.

The exponent of "Deedhiti" (Tattvachintamani) rebutted his dilemma in his own immediate way. He observes that the predicate is inferred on the forte of desire for Inference despite the presence of its previous conviction. This fact stands legalized by the testimony of experience and as such is not liable to be called in question. Now the sum of the negation of those desires with presence of which is invariable followed by the emergence Inference despite the existence of previous conviction is to be taken as criterion of the previous conviction.

And the absence of this previous conviction qualified in the way indicated above – is the essential character of the logical subject. This novel enunciation circumvents the dilemma which has been urged against both the presence and absence of the attitude of desire intended to serve as a stimulant.

Since into the body of this amendment has entered all individual desires in their individual character. So the question of insertion or non-insertion of the aptitude or criterion has been absolutely immaterial.

But this modification, too, is not free from loopholes. Because different types of desire may act as a stimulant with reference to one and identical case of Inference-

Suppose the presence of fire is intended to be inferred in the hill – despite its existence of its antecedent awareness. Now in this given situation the service of desire qua stimulant is felt essential. And it may assume the following varieties:-

(a) Let there be an Inference of fire in the hill.

(b) Let there be an Inference of the predicate other than fire.

(c) Let there be an awareness of the predicate fire et seq.

Here such equation renders it apparent that diversity of desire is infinite. So how can each and every individual desire be comprehended without reference to its common character? Now the question can be answered in the way indicated below-

It has been stated that antecedent conviction of the probandum acts as a deterrent of Inference. So its negation is regarded as the condition precedent of Inference. But it comes out that the negation of the previous conviction has been accorded the status of the character of logical subject. Now it is remarkable that the negation of the previous conviction brings about its result by remaining unknown. Its sheer presence in its own respect is sufficient for the presentation of Inference. It renders its service by halting in cognitio. But the negation of the antecedent cognition taken by itself is not adequate to produce the contemplated Inference.

It is owing to the reason that the service of the desire has been called into requisition to play the role of stimulant. Now return to the above contention – Granted that the variety of stimulating desire is endless. But this fact cannot undermine the causal relationship existing between the character of the logical subject and Inference.

Because the necessity of the knowledge of each and every individual desire has been completely dispensed with. Since the very nature of causal (relationship implies that the negation of the previous requisation accompanied with the sum of negations of desire must be there. But its awareness is not essential. Now when the apprehension of the substantive viz. the negation is not necessary – the question of the awareness of its qualification may be left out of account.

The same argument is applicable to the qualification of the qualifications with equal propriety. Therefore, the conclusion is irresistible that the inception of each and every individual desire without reference to their common character will not be effective of any difficulty.

Neo-Logic Gadadhar Bhattacharya has approached this problem from different angle and offered his solution which, also, carries weight with it. It is attested by the fact of experience that a class of effect is brought about by sharing more than one class of causes.

The individuals which constitute the class cannot be susceptible of delimitation without a clear reference to their common character. And in the event of its absence the causal relation will be infinite. Therefore, specific reference to the common character is felt essential in the matter of the expression of causal relationship. On the background of this digression (Vishayāntara) it may be argued that individual relations of desire are bereft of the required common character. And as its sequel there would be the possibility of the determination of causal relationship. But in the present context this objection is wide of the mark. It is true that the individual negations of desires which are components of the sum – are endless to enumerate. But they have accorded the status of adjective in order to qualify the previous conviction of the probandum. And the negation of the previous conviction as qualified by them has seen the character of the logical subject. It is a plain fact that the negation of the previous conviction which occupies the position of the supreme importance, has a common character to enclose its different varieties within a fold of a single act of cognition. When primary cause has been brought within the compass of the cognitive act, the plurality of the qualification of its qualifications cannot entail infinitude or causal relationship. It is worthy of note that the qualification stands for the previous conviction and its qualifications refer to individual negations of desire for Inference.

There is a little difference between the 02 interpretations –

(a) The former asserts that the logical character is such type of cause of inference that its awareness is not felt essential, and

(b) The latter shows that there is no room for the infinite numbers of causal relationship despite the absence of common character in diverse or discrete negations of desire.

I have shown in the preceding paragraph that the causal relationship between the character of logical subject and the inference cannot be impeached on the score of its infinite variety. But a mature reflection reveals that there is no room for fresh dubiety. The sum of negation of desire has been asserted to be an adjective of the previous conviction of the probandum in the logical subject.

Now the substantive- adjectival relation which holds between the sum of negations and previous conviction may be of 02 types –

(a) The individual negations of desire may be treated as direct qualifications of the previous conviction at one time.

(b) By treating each preceding individual negation as the adjective of the succeeding one in a given series and making the ultimate negation of desire belonging to the self-game series-adjective of previous conviction.

Now 02 competing alternatives are present. But the question preferring one to the exclusion of the other will be ultimately decided by the Law of Parsimony.

It is worthy of remark that the recollection despite its assuming the character of synthetic judgement, is, also, of the nature of the conviction of the probandum. Since it takes stock of the probandum in the logical subject. As a sequel of this recollection is bound to be an impediment to Inference. The fact that has been recurred frequently to an antecedent awareness of the probandum intervenes as a prevention of Inference. Therefore, it is apparent that the presence of the recollection qua synthetic judgment is bound to remain abortive without desire for Inference as an incentive factor.

But it is equally manifest that the contemplated Inference is the wherewithal to the achievement of a desired objective. It is owing to this palpable reason that this recollection has been asserted as embodying that fact viz. Inference of the probandum and its sub-concomitant are instrumental to the attainment of the cherished end. In this given situation there takes place the desire for inference of the probandum and its sub-concomitant immediately after the recollection in question. The presence of the desire is followed by the drawing of inference – the logical subject holds the probandum and its sub-concomitant.

However, a fresh poser springs up by itself-

The present inference owns all essential characters of the synthetic judgement qua condition of inference of the probandum in the logical subject. Because the inference takes cognizance of the sub-concomitant of the probandum in the logical subject.

To be more explicit – This inference has the sub-concomitant of the probandum as its objective datum. Over and above the sub-concomitant stands cognized as existent in the logical subject.

If so, why the **2nd Inference** of the probandum will not recur consequently for the 2nd time on the basis of the previous inference qua synthetic judgement? It may be argued that the inference qua synthetic judgement in question is of the nature of the conviction of the probandum. Since it is conversant with the probandum in the logical subject. And so it will act as a bar to the entailment of the **2nd Inference**. But this defence is devoid of solidity. Because the desire which preceded the succession of the **1st Inference** is there. It is on the basis of this immediately antecedent desire that the **1st Inference** comes into existence. Now the desire lasts as a stimulant upto the moment of emergence of the inference qua synthetic judgement. And this desire undermines the obstructive wherewithal or forte of the conviction of the probandum qua inference and as such there remains nothing to prevent the occurrence of the **2nd Inference.** So the charge of the undesirable entailment of the **2nd Inference** in succession remains unanswered.

A section of logicians id inclined to hold the **1st supposition** is worthy of acceptance. Since the **2nd one** is infected by ado of thought in combrousness.

Because it has been held that one individual negation of desire would be the objective of the other. And the same process will be repeated with respect to the rest of the series in order to link them together.

Now what is intended to be an adjective will require the assumption of relation with its substantive negation. So an individual negation would be treated as an adjective of another negation. And the substantive negation would equally be considered as an adjective of the **3rd negation**. The process will be extended up to the end of the series. As a result of it, an admission of the endless numbers of distinct relations between all adjective – negation + substantive – negation would be inevitable.

Such event accounts for coinciding preferential treatment to the 1st alternative referred to above. A group of logicians contends that the defect which taints the succeeding chance is equally present in the proceeding one.

Because desire-negations which have entered into its constituents need to be invested with the status of any adjective of the previous conviction. And consequently such event will entail the admission of infinite numbers of substantive – adjectival types of relations. So the prolixity of assuming an endless series of relations is shared by the both alternatives in an equal degree. There is no relational ground for extending the seal of approval to the **1st hypothesis.**

Such logical subject or knowledge-base is of the following kinds –

(a) Paksha Hetu (Middle term) – logical reason for inference) – The hill holds fire-where the probandum is inferred with the logical subject.

(b) Sapaksha Hetu (containing reason-the subject of Inference) – Here antecedently existence of the probandum in the logical subject – the smoke probans already existed with the wet-wooden fire in the wood – lit kitchen.

(c) Vipaksha Hetu (absent or negative reason in Major term) – As there is the negation of fire in water.

(d) Asatpratipaksha Hetu (non-existence of else parallel reason) – where there is no anti-parallel else forcing reason due to invariable concomitance as the adjectival-logical subject-as-there is no other causing ground of Inference of having fire except smoke. Smoke-cause is the vital event.

(e) Abādhit Hetu (non-refutative reason) – the absent character of the probandum is not determined in the logical subject as the absence of fire is not determined by any other evidences.

The Nature and Utility of Synthetic Judgement (Parāmarsha)

I propose to appropriate the present chapter to an explosition of the essential nature of synthetic judgement. I also intend to examine the legitimacy of its claim to be considered as a vital condition of Inference hinging on 03 conditions:-

(a) The knowledge of the universal concomitance for the probans and the probandum.

(b) The knowledge of the existence of the probans in the logical subject.

(c) The knowledge of that logical subject is possessed of the probans as the sub-concomitant of the probandum.

The last condition passes under the name of synthetic judgement. Since the judgement in question comprises of 02 distinct aspects of the probans –

(1) Existence of the logical subject and its universal concomitance

(2) The probandum

Now the vital role of synthetic judgement is apparent from the fact that it immediately precedes the materialization of inferential judgement as a class. It is owing to this reason that synthetic judgement has been incorporated into the body of the definition of Inference.

The Neo-logician maintains that synthetic judgement being conversant with the existence of the probans with its concomitance and the probandum are self-same and unique in essential nature. It preserves unity without occasioning any mismatch. It becomes the immediate condition of inference while maintaining inner unity.

So far the judgement per se is taken into account, it owns oneness like other accredited judgement based on the relation substantive-cum-adjectival. The logicians postulate that the synthetic judgement qua condition of Inference is the quality of the Self and as such resides in it through the relation of inherence (Samavāya). And the inference takes place in that Self in which there are observation of the probans and the knowledge of universal concomitance by the relation of inherence

(Samavāya). A subtle analysis of its nature unfolds that it is conservant with 03 objectives viz.

(1) Universal concomitance (Vyāpti)

(2) The probans (Dhuma)

(3) Logical subject (pakshata)

There exists the relation of substantive-cum-adjectival type inter se. That is to say, universal concomitance is the objective of the probans and the probans is that of the logical subject. So this synthetic judgement is susceptible of 03 modes of expression:-

(a) It is the judgement that serves to indicate the universal concomitance – as an adjective

(b) The probans – as an adjective

(c) The subject as – a substantive

When attempts made to link them together through the relation of substantive-adjectival, it faces the charge of the absence of crucial evidence (Vinigaman Virah). And this would ultimately entail varieties of causal relationship between Inference and synthetic judgement – Suppose:-

A, B, and C, are mutually related through substantive- adjectival relation.

Again suppose that there is no rigid rule as to which of the 03 should be invested with the status of an adjective.

Here when there is no decisive evidence (Vinigaman) in favour of any of them, each and every individual deserves to be accounted as an adjective by rotation.

To be more elaborate –

'A' is an adjective to 'B' which is also an adjective to 'C'

'C' is an adjective to 'B' which is also an adjective to 'A'

'B' is an adjective to 'A' which is also an adjective to 'C'

'B' is an adjective to 'C' which is also an adjective to 'A'

'C' is an adjective to 'A' which is also an adjective to 'B'

'A' is an adjective to 'C' which is also an adjective to 'B'

So the admission of synthetic judgement entails firing thoughts. Here the smoke qua sub-concomitance of fire in the hill is in contact with the organ of vision.

The recollection of sub-concomitant of the probandum (fire) is also there. The class-character smokehood with is the determinant of universal concomitance (Vyaptavachhedak) is also shared by the smoke under actual observation. Consequently, it takes place of the cognition of identity between the sub-concomitant and the smoke in question. Now the answer may be said down:-

(a) Repeated observation of the link between 02 phenomena having 02 distinctive individuals.

(b) Absence of the observation of any deviation from this uniformly recurring events. The ultimate result confirms the classknowledge of concomitance as the ultimate foundation of inference. This classcharacter occupies the position of determinant (Avacchedaka). Now take for instance- the case of smoke and fire. Due to spatial limitation smoke entitled-subconcomitant. Where – fire on account of its spatial extension is called superconcomitant. Here prevails the class-character of the determined (Avacchedya),that is, of the thing to which it is determinant of synthetic judgement.

Therefore, it is evident that the union of the conditions entails its affirmation. So there is the justification to admit it as an immediate condition of inference.

An analogous case is being furnished by recognition-(pratyāvigya)

Now the object qualifies by thisness (this smoke) in contact with the visual organ. And there is the recollection of the object circumscribed by thatness (this is that smoke) as a matter of fact. So recognition as such embodies to identify the object (smoke) referred to by this and that respectively. So synthetic judgement, also, stands on equal footing with recognition or perceptual cognition of the smoke and recollection of the universal concomitance which constitute essential nature individually. The observation of smoke and recollection of universal concomitance are the accredited condition of inference. Here the conditions of Inference and that of perceptual cognition are equally present in the case in hand. Now crucial problem burst forth –

Is the assemblage or blend of conditions of perceptual cognition is superior to the assemblage or blend of the conditions of Inference or vice versa?

Here if the objects of the 02 types of cognition are identical, the blend of the conditions of perceptual cognition overrides that of Inference one.

But this position is reversible in the event of the diversity of their objects. Since in that situation the assemblage of conditions of inferential cognition cancels that of the perceptual one.

As logicians admit the causal system to investigate legal dispute, perceptual cognition is susceptible of 02 fold classifications viz.

(a) ordinary or common customs (Laukika)

(b) extraordinary or uncommon customs (Alaukika)

Here so far the 1st type of cognition is taken into account, the question of superiority is an universally avowed fact. So it is only between the assemblage of the conditions of uncommon perceptual cognition and inference that the claim to superiority remains confined. Now a question may crop up as to what is the reason to assert that the assemblage of the conditions of inferential cognition prevails over of the perceptual one in objective difference. Due to this reason, the inferential cognition is based on the relation of inherence as we observe water (qualified) due to waterhood (qualifier). Whereas we involve in delusive knowledge in mirage due to the absent of the relation of inherence between sunrays and water. So to clarify the whole situation-synthetic judgement is a case of determinate cognition (Vyavsaya). Consequently, there is the assemblage of the conditions of its introspection or consciousness of perception. So the non-admission of superiority of the assemblage of the condition of inferential cognition would be suicidal to its very independent existence. Since there would be the introspection of the synthetic judgement due to the assemblage of its conditions. And this would ultimately entail an ugly consequence in the forfeiture of the affairs of Inference. This is one of the basic reasons for admitting the superiority of the assemblage of the conditions of Inference.

There is another consideration of equal importance. The question of superiority is closely associated with the natural disposition of human-determinate mind as its component. Here the absolute superiority of the assemblage of conditions of ordinary perceptual cognition over others cannot be called in question. But the point that deserves utmost consideration is – which of the assemblage of conditions presupposes greater human effort for its realization?

Now it stands to reason that the assemblage of conditions requiring greater degree of effort sets aside lesser amount of exertion for its achievement. And this accounts for according superior status of the hodge-podge condition of inference to that of extraordinary perceptual cognition. Here the identify or the recognition is the subject of extraordinary perceptual cognition. Since all normal cases of perceptual cognition hinge on the immediate connection of the sense-organ with the object (smoke) in the logical subject. But the smoke which was antecedently determined as the sub-concomitant of fire remains beyond the range of the organ of vision.

Therefore a chain of relation has been hypothesized in order to bring sub-concomitant of fire into relation with the smoke. And that is technically called "cognition-qua-contact" (Sansarga gyan) by the logicians.

Here successive link moves as –

(a) The visual organ is in contact with the mind

(b) The mind in turn is in contact with Soul in which dwells recollection of the sub-concomitant of fire through the relation of inherence (Samvāya).

Here the logicians convey that the recollection permits its object to be the object of visual perception. As such recollection is accorded that rank of contact in the present situation. Contact as such presupposes the existence of 04 terms:-

(i) Visual organ

(ii) Cognition contact

(iii) Recollection

(iv) Synthetic judgement

Here unless the visual organ is linked with recollection, recollection will cease to be in contact of the organ of vision. So a series of relation occurs to invest recollection with status of contact between visual organ and the object of recollection.

In the 1st instance there arises the knowledge of universal concomitance of smoke and fire with reference to their class-character-smokehood and firehood respectively. In the 2nd instance there arises the synthetic judgement.

The hill holds **smoke-qua-sub-concomitant of fire**. In the given situation – the smoke which observed in the hill lies in an ordinary contact with the visual organ. And as such it becomes an object of synthetic judgement. But the awareness of identity of the smoke antecedently determined (Avacchedya) as the sub-concomitant of fire, takes place on the basis of **recollection-qua-extraordinary contact.**

Besides, the observation of smokehood which is determinant of universal concomitance also shared in the production of consciousness of identity. Now the observation of smoke in the hill and the knowledge of the universal concomitance of smoke + fire that are accredited conditions of Inference assembled there. Consequently, this would supersede the assemblage of conditions of synthetic judgement as counter-posed by the Mimansists.

In order to operate upon the urge of Mimansists, logicians deal the situation in other way:-

Suppose, there is the recollection of universal concomitance and smoke is the sub-concomitant of fire. This is engendered by a condition other than the smoke which is in contact with the visual organ. And there would occur the perceptual cognition of the smoke in question. Here it is worthy of note that the perceptual cognition is not confined to pure smoke and as such the smoke being one of the terms of the relation of universal concomitance conjoins smoke and fire. So far the perceptual cognition of smoke is taken into account, it is ordinary in its nature due to its **origination** from an ordinary contact. But the position is entirely reversible in the case of universal concomitance. It cannot be other than as extra-ordinary perceptual cognition of which the universal concomitance ought to be expected as an object. In this unitary cognition smoke by itself makes felt in 02 distinct aspects:-

(i) It is existent in the logical subject and

(ii) It is sub-concomitant of fire.

So it is entitled to be called a synthetic judgement.

The net result is that this is not liable to be set aside by the collocation of the conditions of Inference. And this impels the affirmation of synthetic judgement as an essential condition of Inference under the Law of Simplicity (Lāghav-occam's razor). But the mimansists object the Law of Simplicity infavour of the antecedent existence of the synthetic judgement in a peculiar situation. They prefer 02 distinct events of Inference:-

(1) Invariable antecedent

(2) The negation of the chance of its being otherwise realized with reference to all conceivable cases of inferential cognition alleged as the Law of Complexity (Gaurava). But the assumption of the element of the negation of being otherwise realized is equally shared in both counter-claims. Both of them suffer drawbacks of one or the other as such are equally debarred from taking recourse to the plea of simplicity + complexity. Further the logician urges a deeper probe of the situation.

A cause to be considered as such must present by itself in its distinct character. This treats as a mark of identification as is technically termed determinant (Avachhedaka).

Now when the question of relative simplicity of the determinant of causal relationship is taken into consideration, the contention of the logician is bound to appear in an advantageous position. 02 distinct types of essential attributes (by Mimansists) that reside in them separately, will be treated as the 02 determinants of the fact of being casuality (Kārantā). That is to say, the individual causes involve 02 distinctive determinants in the shape of 02 essential attributes existing in them. Further the 02 cognitions are-

(i) The knowledge of the probans as the sub-concomitant of the probandum

(ii) The observation of the probans in the logical subject as universal concomitance – which occupies the position of cause. So the objects with which the cognition is conversant are regarded as the determinant (Avacchedaka). It is worthy of note that these objects are associated together through the substantive-adjectival type of relation. The sum and substance is that the probans has been the object of the 02 distinct cognitions and thus making itself felt as 02 separate determinants of the 02 distinct causalities. So it is evident that being one or single instance of cause, the probans (smoke) is bound to present by itself 02 times as distinct determinants due to the difference in 02 **cognition-qua-causes.** What is intended to convey is that smokehood (class-character) which resides in the probans (smoke), makes itself felt as the determinant of the sub-concomitance.

The knowledge of universal concomitance which refers to smokehood as determinant of the fact of sub-concomitance (Vyapyatā Vachhedaka) is asserted by the Mimansist as a separate cause of Inference. Here universal concomitance is of 02 termed.

So its knowledge invariably presupposes the knowledge of them (02 termed).

It is an avowed fact that relation cannot be assumed as detached from its related terms. Whenever the relation of universal concomitance takes place referring to 02 termed-

(i) Super Sub-concomitant (Vyapaka)

(ii) Sub-concomitant (Vyapya)

Here the existence of the former is sought to be proved in the subject and as such it is accorded the status of probandum. And the latter serves as a ground of Inference. of the existence of the probandum and as such occupies the place of probans. The net result is that the knowledge of universal concomitance is bound to involve the knowledge of a sub-concomitant-smoke. As it immediate corollary, the generic attribute (smokehood) that dwells in the sub-concomitant (smoke) is regarded as the determinant of sub-concomitance.

To be more clear –

(a) The universal concomitance

(b) Its 02 terms

(c) Their generic attributes – are the objects of the self-same knowledge of universal concomitance.

Here the Mimansist asserts that the knowledge of the existence of the probans in the logical subject is also a separate condition of inference. But the logician holds synthetic judgement as its cause in lieu of the 02 distinct ones as advocated by the Mimansist. But his isolated causes of inference may design absurdity of thought.

So the logician admits synthetic judgement as the only immediate cause of Inference. It is owing to this reason the logician holds that synthetic judgement must own the following features to be regarded as the cause of inferential judgement –

(i) It must refer to the relation of the probans with logical subject.

(ii) It ought to refer the self-same probans as qualified by universal concomitance. And this has been technically called **the knowledge of the relation of related facts as qualified by the universal concomitance (Vyapti Vishishta Vaishishtya-vahigyan).** And this renders it evident that there is no need to refer to smokehood as the determinant of sub-concomitance in order to invest with the position of cause. So the presence of the cause should be invariably followed by that of its effect under the Law of Causation.

Further the absence of the former entails the absence of the latter. As the situation stands, the occurrence of the Inference is felt eventually despite the absence of the synthetic jugdement. So insertion of the condition of synthetic judgement in the affairs of inference may violate the Law of Causation posed by the Mimansist in favour of 02 conditions viz.

(i) The knowledge of the probans

(ii) Recollection of the universal concomitance

To evade such incongruity let me argue-

Suppose, the visual organ is in contact with smoke issuing from the fire place. This contact is not limited to the individual smoke but embraces smokehood as such. And the universal character of this smokehood is not liable to dispute.

It inheres all individual smokes including those existing in the hill. Consequently it turns out the smokehood which dwells in smoke of the hill is equally in contact with visual organ. Here ordinary contact causes the conditional for inference.

Now consider the fact in otherway-

The smoke and fire stand related in the shape of sub-concomitant and super sub-concomitant- is a known fact. And it is equally deducible that smokehood residing in each and every individual smoke is the sub-concomitant of the universal concomitance of smoke and fire.

Where there is smokehood there is universal concomitance of the 02 terms referred to above. As a result of which- so far as smokehood is concerned, there is no ordinary contact with organ of vision. And there is extraordinary contact with the universal concomitance of smoke and fire. It is worthy of note that these 02 contacts have engendered the synthetic judgement in which smokehood that exists in the individual smoke of the hill, stands as substantive and the universal concomitance of smoke and fire occupies the position of adjective equivalent to the synthetic judgement. This would ease the occurrence of Inference – that the smoke in the hill is the sub-concomitant of the universal concomitance having its referent. So this Inference of the universal concomitance is the nature of synthetic judgement. Now when the synthetic judgement is there, there is nothing to debar the drawing of Inference even on the strength of the probans that remains beyond the ambit of contact with visual organs.

But it is an undeniable fact that the knowledge of the smoke received by an authentic organ of cognition would ease the recollection of universal concomitance of smoke and fire. And ultimately the mind by receiving its cooperation as an auxiliary factor will cause synthetic jugdement in normal affairs. So the causal relation between synthetic judgement and inference will remain intact.

Now synthetic judgement is sui generis in its nature. It ought to be considered as the product of an organ of cognition that can be cased under its accredited varieties. The Neo-logician pleads that the smoke is the sub-concomitant of fire being conventional one. It stands to reason that this cannot take place in the event of the absence of specific judgement. Here the mind is competent to produce the cognition of the objective datum by turning upon the exclusive service of the extra-ordinary contact to realize judgement being responsible.

Against the Mimansist's assumption of the knowledge of the determinant of the sub-concomitant as conditional adjective of Inference, the logician asserts – the hill holds the sub-concomitant of fire as the genuine case of synthetic judgement based on trustworthy verbal testimony of the asserter. Here Inference is –characterized by the absence of the knowledge of the determinant of the sub-concomitant of fire as opposed to the condition of Inference.

Again suppose – a person deserves smoke in the hill and asserts – there is smoke.

The knowledge of the object-smoke makes itself felt together with its property smokehood (character). The class-character (smokehood) can only exist in the smoke. So it is apparent that smokehood as such cannot be regarded as super concomitant of the existence in the hill. Here the case cannot be asserted –

Where there is the existence in the hill (Parvatvrittitva), there is smokehood.

This knowledge of smokehood is to be understood with respect to something that exists in the hill. This something is no other than smoke as such.

Here in the given situation, a dubiety occurs in the mind – is smoke in the hill concomitant of fire or not?

This would entail the emergence of Inference inevitable in spite of dubiety regarding the fundamental principle of Logic.

Here the recollection (smoke) being the concomitant of fire is the condition precedent of inference of fire. Its previous existence of the cognitive relationship in mind can impede the occurrence of dubiety. Here the recollection of the universal sub-concomitance finds its expression through the medium of judgement – smoke as the sub-concomitant of fire. Since the cognition of smokehood (adjective) in the smoke on the hill will thwart the emergence of dubiety on basic principle-

When in a locus possessed of characteristic property, there occurs the cognition – the cognition of the excluding property – a dubiety cannot take place with respect to the property that has been excluded.

A concrete instance may be taken to test its applicability – dubiety – is it a tree stump or not is an avowed one. Its analysis exhibits that this stands for the locus or subject. And both of the fact being a tree stump and the negation of the fact of being a tree stump are the 02 properties possessed of. Now suppose there takes place the cognition of the negation of the fact of being a tree stump in human beings.

Now in this given situation there cannot arise "dubiety" – Is this human being as tree-stump or not?

This instance renders it apparent that the cognition of the negation of the fact of being a tree stump has taken place in that locus which is characterized by the property of the fact of being a human being.

The former property excludes the property viz. the fact of being a tree stump by its very existence.

To be more accurate – these 02 properties mutually exclusive. The presence of one implies the negation of the other and vice versa. This stands for human being. So "that dubiety" of the fact of being a tree stump cannot find a room for its appearance.

A classification of excluding factor may be set down in the following order-

(i) The negation of an entity is in question. Negation of 'A' may be regarded as its excluding factor. That is to say, the negation of 'A' implies the exclusion of its possible existence.

(ii) The sub-concomitant of the negation of an entity (Tadbhāvvyāpya) considers the stock in trade example of dubiety. The sub-concomitant of the fact of being a tree stump occurs its leafy branches. And that of the fact of being a human being are hands and feet. Now hands and feet are the sub-concomitant of the negation of the fact being tree stump.

To be more accurate, hands and feet are the sub-concomitant of both of them. Now the negation of them (hands & feet) will exclude the cognition of the fact of being a tree stump. As hands and feet are the excluding factors, their cognition eliminates the chance of the cognition of tree-stump. And consequently this excludes the dubiety of tree-stump in the case under review.

(iii) That property which exists in a locus with reference to them-there takes place the cognition of the negation of an entity. The cognition of negation implies its locus. And it is apparent that this locus is equally recognized like the negation in the case of the cognition of negation.

And this locus makes itself felt as possessed of a particular property. And the result is that this property serves as an index of negation that shares its same locus.

Here it is reasonable to assert that the index of negation will shift the centre of weight upon the negation as such to the exclusion of that what has been negated or its counter-term (pratiyogi). This is a self-evident fact not liable to dispute. So it is but legitimate to expect that its presence will exclude that cognition of which has been excluded by it.

Let me consider – how far the classical example of dubiety stands?

In this predicament that "dubiety" in question cannot take place in any conceivable manner. Here the dubiety alluded to above cannot exist in the mind. So it is apparent that the function of exclusion that belongs to the other 02 varieties of excluding property is equally shared by the excluding property under review.

Further the negation that operates as an excluding factor is of absolute type (Atyantabhāva). A reciprocal negation of the container and the absolute negation of its content are co-extensive in their essential nature.

Let me propose a concrete instance – take for the absolute negation of the table on the ground. In the same situation, the ground cannot be regarded as the container of the table. Consequently, there is the reciprocal negation of its container in the ground. So ultimately the case turns out where there is the absolute negation of the content, there is also the reciprocal negation of its container.

Similarly, where there is the negation of the absolute negation of the content, there is also the negation of the reciprocal negation of the container. It is worthy of note that the negation of the negation arithmetically implies the positive sense.

Therefore, the assertion of the negation of the absolute negation of the content is tantamount to the assertion of the opposite existence on the content. Consequently there cannot be the reciprocal negation of the container due to the presence of the content. That is to say – the reciprocal negation of the container will exclude the chance of the existence of the content. So it is obvious that so far the function of the exclusion is taken into account –both the absolute and the reciprocal negations stand on an equal footing.

To express in terms of the Neo-logician-eitherness (Anyataratva) of the positive or the negation may be regarded as efficient to discharge the function of the probans. If so, the intended inference will find no difficulty to take its rise.

Here the logical significance of eitherness may be classified as the word "either" signifies one or the other of the 02. Here this conjunction means alternative. The word alternative as such embraces the 02 identifiers in any equal degree. There is no question of exclusion of the one by the other. It is apparent that a definite connotation (Pravrittinimitti) must be attributed to the word "either" in order to enable it to connote the attribute possessed by its meaning. Consequently the attribute "eitherness" which is normally expected to be connoted by it deserves a subtle analysis – that which is different from that things which are different from the each of the 02 viz- from the both receives the designation of either.

Let me narrate – either 'A' or 'B', now 'C' and the entire universe as such are different from each of the 02 viz. 'A' and 'B'.

To put it in a plain way –excepting 'A' and 'B' the entire universe is different from 'A' and 'B'. Here it stands to reason that 'A' and 'B' cannot differ from 'A' and 'B' owing to their self-identity.

'A' is always 'A'

'B' is always 'B'

And as much 'A' cannot differ from 'A'

And 'B' cannot differ from 'B'

Now the difference of 'A' and 'B' exists in 'C'

And the entire universe and the difference of them exist in 'A' and 'B'.

Here the 2nd difference exists in each of the 02 viz. in either 'A' or 'B' taken in their individual status.

The 2nd difference is identical with eitherness. So to sum up the difference of the difference is identical with eitherness. Now in the light of dilation it is evident that the assertion of difference from the different of the positive or the negative entity is tantamount to the assertion eitherness of the positive or negative entity.

However the contemplated inference resorting to eitherness of positive or negative entity as the probans fails to deliver the desired results.

Since the probans is undependable on the score of its impossibility, all the entities of the universe are susceptible of classification under 02 heads –

(i) Positive

(ii) Negative

Besides positive and negative entities, there is not the remotest possibility of a 3rd variety of entity that may be regarded as distinct from them. So it is evident that eitherness of the positive and negative entities cannot play the role of a valid probans to its inherent weakness.

Again since eitherness of positiveness or negativeness cannot be impeached on the alleged footing. Innumerable properties are different from both of them as substancehood, earthhood and the like. These properties are determinate concomitant (Vyāpya) of positiveness. So it is other than its determinant concomitance (Vyapaka) namely-positiveness. It is too obvious to need to mention that is different from negativeness. So the legitimacy of this type of probans cannot be liable to be called in question. In counter disclaim by the Mimansist in the concept of universal subsistence for exclusively affirmative Inference (Kevalānvaya), the Neo-logician has modified his plea of the justification of synthetic judgement– when the cognition of the negation as a thing which operates as a bar to the determination, the determination of the former (thing) is a condition of the latter (determination). A concrete instance will bear out the truth of this cardinal canon-

If there be the cognition of negation of smoke as such, the inferential judgement of fire is bound to remain unrealized. So the determination of the smoke-qua-probans deserves to be accounted as the condition of Inference in question.

Now suppose – there is the cognition of the probans – smoke in the logical subject namely – the hill. So it is an undeniable fact that this smoke is the sub-concomitant of fire.

Again suppose – there is the cognition (smoke) which is different from the sub-concomitant of fire. In this given situation the Inference of fire does not take place to its rise. Therefore, in the light of the alleged logical theory of the determination, the probans is identical with the sub-concomitant of the probandum as the condition of Inference. And this cognitive identity is equivalent to synthetic judgement.

As discussed above, smokehood is evidently discernible in both (recollection of universal concomitance of smoke with fire and the cognition of the existence of smoke in the logical subject – hill) conditions of inference. Consequently its cognition is equivalent to the cognition of specific property (vishesha Dharma). It acts as a determinant of the universal concomitance. So the cognition of smokehood will not suffer the emergence of the apprehension of differenc of sub-concomitant (Vyapyabheda).

On the contrary Inference will take place as normal affairs due to the assemblage of its conditions. So the impossibility of the apprehension of the difference of sub-concomitant denies its right to be considered as an impediment. And this impossibility of the impediment proves fatal to the principle upon which the logician has pinned his faith to justify the utility of synthetic judgement.

The logician avers that the cognition of probans existing in the logical subject as identical with the sub-concomitant of the probandum stands in a advantageous position since the cognition of identity has simplicity to its credit. Therefore the claim of synthetic judgement cannot be dismissed without committing violence to logical sense.

Here the proponent and the opponent (Neo-logician and Mimansist) concur that the negation of the cognition of difference precedes all cases of Inference. Consequently this provides a rational ground to hold it as the cause of Inference of simplicity and complexity of thought upon which reliance has been placed by the Neo-logician in order to preferential treatment to synthetic judgement.

Further the presence of the cognition of difference would be fatal to synthetic judgement. A question may crop up-when the precedential existence of cause and succession of effect is an avowed fact in logical tenets. So synthetic judgement and its cause cannot make their appearance in an identical unit of time. The impossibility of co-evality of them places the question of simplicity and complexity beyond the bound of consideration.

As the Mimansist disagrees to incorporate synthetic judgement as the condition of inference, he asserts only 02 distinct and independent cognitions in the affairs of inference viz.

(i) the recollection of universal concomitance

(ii) the cognition having the determinant of the sub-concomitant smokehood as an adjective or qualifying attribute to its productives.

Here the admission of the cognition of difference as an impediment would entail the admission of synthetic judgement as the condition of inference. Since it is in this eventuality, the cognition of difference can be expected to be in opposition to the originating cognition and thus enabling it to operate as an impediment. And its ultimate result would be synthetic judgement accounted for the condition of inference. The cognition of the probans is the condition of synthetic judgement. The observation of smoke in the logical subject induces the recollection of universal concomitance of smoke and fire. And this ultimately occasions synthetic judgement.

A cognition must have its utility in the matter of the production of synthetic judgement. So it is bound to produce a certificate of its usefulness in order to validate its claim to be considered as such.

It has already been stated that the cognition of the probans produces the recollection of universal concomitance which is the nature of a relation between 02 terms, says the probans and the probandum.

As 02 termed relation – the cognition of 01 term occasions the recollection of the other (02) term. It is worthy of note that the cognition of the term involves the cognition of property that dwells in it. That is to say, both of them are the contents of the self-same cognition. The reason lies in the fact that this property in question functions as determinant of the relation subsisting between them.

To be more clear, the cognition of smoke-qua-probans takes place together with its property (Dharma) viz. smokehood as its content. Non-compliance with the condition of universal concomitant of smoke and fire is logically viewed as determinated (Avachhedya) by its determinant (Avachhedaka – smokehood).

Now in the case of recollection of universal concomitance by means of cognition of smoke-qua-probans, it is essential that smokehood also manifests itself as the contact of the smoke cognition.

If the smokehood fails to appear as such, the cognition of the smoke would not be in a position to render any service to the emergence of synthetic judgement.

The simple awareness of the probans is incompetent to serve any useful purpose of the synthetic judgement. So it is essential that its cognition should be conversant with its property (Dharma) that operates as the determinant of the universal concomitance.

Now the common property (cowhood) is equally owned by all individual cows. Experience attests the fact that there is common and essential attribute which belongs to every member of a class. A rational mind recognizes different individual or things and ultimately discovers that they share an essential property in common. This ultimately is responsible for the production of a general idea in shape of cowhood and the like. It is due to the residence of this property that the observation if any individual produces the cognition – this is a cow.

All the members constituting a group differ from one another is shape, colour, structure and the like. But the similar property in cognition is a felt fact to recognize cows at any place despite the fact that they differ from one another. Consequently there remains only the common property to account for this common character of cognition. And this provides the ground for assuming the existence of generic property or a class-character. This generic property ought to abide in all members constituting a group and remains isolated from other animals. So the fundamental nature of generic property has been rendered crystal clear.

In a syllogism- the hill holds fire because blue smoke is emitting out of it.

Here the property "blueness" or "blackness" or the like in the smoke may have endess properties in smoke causing complexity of observation. And blue smokehood or black smokehood cannot the be determinant property of sub-concomitance due to non-compliance of the terms and conditions of the logical – base enunciated above. Here the Law of parsimony (Lāghava) cleanses the issue in favour of smokehood (intrinsic character). When simple property is competent to occupy the position of determinant, the combrous property ought to be left out of consideration.

In other way, the charge against complexity or over extension is justified on the inorganic properties of smoke instead of its organic property-smokehood. The portional existence of variable characters in all smokes is vitiated by the discursive spinner.

Inference as a class is founded upon valid probans. But the probans per se cannot discharge its function unless it is recognized with reference to its relation. The probans and the probandum are regarded as the sub-concomitant and the super-sub-concomitant respectively. But this requires the postulation of a relation by means of which they may be entitled to be termed as such:

If the relation were not rigorously referred to, the probans as a class would be non-concomitant of the probandum. In the present context, non-adherence to the relation would prove fatal to the claim of cowhood as a genuine probans despite the reference to the 02 components of cowhood. Temporal relation in a particular moment supersedes the relation of inherence as the determinant of probanshood. Here the fact of being connotation of the word "cow" has been asserted as the probandum. And cowhood has invested with the status of the probans by referring to its relation of identity (Tadātmya) as the determinant relation of probanshood.

Here the observation of the specific characteristic of one of the alternatives of dubiety (eitherness-smoke or light-fire) serves as a cooperative agent of the perception-qua-organ of cognition. The cognition of the sub-concomitant property of a thing is in direct opposition to the cognition of the negation of the object in question. It has been shown that the observation of the rule-concomitant property is equivalent to that of specific character. In this light, the cognition eitherness would serve to exclude the possibility of cognition of its opposite entity. Since the joined method of argument and difference (Anvyavyatireki) affords testimony being dilated in inferential devices:

Here the universal concomitance is validated by at least 03 methods as per perception of relation between smoke and fire in the wet-wood flaming kitchen qualified by pervasive concomitance-

(i) Kevalānvayi (exclusively logical connection) between hetu (Middle term cause) and Sādhya (Major term-effect)

Where there is smoke there is fire (Tatsatve Tatsatvam)

(ii) Kevalavyatireki (exclusively non-existent or negation) where there is no smoke, there is no fire- Tadabhāve Tadabhāvah)

(iii) Anvyavyatireki (positive and negative assertion) –where a critical or comparative study to adjudge similitude ($x = y$) and dissimititude ($x \neq y$) in the knowledge –base (Pakshatā) to confer valid conclusion on prior perception, testimony and invariable concomitance.

Such universal concomitance or pervasion being the nerve of inference is of 02 types-

(1) Samvyapti (equipotential or tantamount concomitance)- Paramount vs tantamount adjective –Nameable or knowable –we infer either of the term from the other as –

What is nameable – is knowable

What is knowable – is nameable

(2) Asamvyapti – (non-equipotential concomitance between the 02) here we infer the term from the other but not vice versa –

Exp – we infer "fire" from smoke but

not infer smoke from fire.

Now reverting to the sequence – it has been repeatedly observed that universal concomitance implies the existence of the probas and the probandum in the self-same locus. It is needless to observe that the fact of co-existence in an identical substratum would be different due to the difference of each as well as identical probans.

To take up a concrete instance - there are innumerable things which entitled to be regarded as the sub-concomitant of fire. But all the cases of universal concomitance are individually different from one another may be brought under a common concept (Anugat pratiti) by the expression-fire concomitance (Vahnivyapti).

And this renders it apparent that eitherness is not an extralimital Anatirikti Vritti) with reference to concomitance. The question of complexity or simplicity is immaterial in the present context. Since an object whether simple or cumbrous is equally entitled to be considered as its non-extralimital. If so, even a cumbrous property would counter with no trouble to be determinant of sub-concomitance. So the view of the Mimansist may be equally worthy of acceptance. But the logician rejoins that the contention is devoid of substance. Since it is undeniable to note that inference takes place even when there is the verbal judgement – this is possessed of the sub-concomitant of fire (Vahnivyapya).

In the case, there is only a reference to the sub-concomitant of fire. Consequently it is evident that the determinant of sub-concomitance is not the object of knowledge in question. So it ought to be admitted that in the present case-inference has been drawn on the basis of the knowledge of universal concomitance. And this clearly evinces that Mimansist's reliance upon the knowledge of the determinant of sub-concomitance as conditio sine qua-non is somewhat irrelevant. This deviation proves fatal to the assumption of the relation of cause and effect between the knowledge of the determinant of sub-concomitant and fire.

Further the awareness –where there is the sub-concomitant of fire, there is fire-entails admission that the invariable concomitance of fire is the determinant of itself. It has been urged that if an object were regarded as the determinant by itself, this would involve the logical defeat styled as self-dependence (Ātmāshraya). The idea of determination implies duality or plurality in all cases without exception. The object to be determined and its determinant cannot be regarded as numerically identical in any conceivable manner. So the admission that the self-same entity is the determinant of itself is bound to face the alleged objection.

Here the Mimansist defends that the individual cases of universal concomitance differ from one another (fire and smoke). Truly there is invariable concomitance of smoke and fire as a class. So it turns out that there are 02 varieties of concomitance viz-

(a) individual concomitance (Vishesha Vyapti)

(b) universal concomitance (Sāmānya Vyāpti)

And this difference permits the one to be treated as determined (Avachhedya) and the other as determinant (Avachhedaka). Here individual concomitance is adequate to meet all requirements. So the assumption of universal concomitance ought to be considered as a mere superfluity. But this is unentertainable. There may be determinanthood in the shape of the negation of the fact of existence in additional places which are devoid of the thing intended to be considered as the non-extralimital itself. Thus makes his (Mimansist's) position unshaken. He affirms that the knowledge having the determinant of sub-concomitance as an adjective – is the condition of Inference. But its utility does not lie in its mere factual presence pleaded by the logician. Were it so, this would entail as absurd consequence.

Suppose (1) a person is devoid of the knowledge that smokehood is the determinant of sub-concomitance of fire.

Again (2) he is in possession of the knowledge, this is possessed of smoke.

In this situation he would be entitled to infer the existence of fire due to the presence of knowledge. Smokehood is the actual determinant of the sub-concomitance of fire.

So in order to plug this loophole, it is essential to make an amended enunciation. There is the property which appears as the determinant of the sub-concomitance of the probandum in the knowledge of the universal concomitance. And the knowledge of the sub-same property as an adjective should be regarded as the condition of Inference. In this case, the knowledge of universal concomitance has been solely provided by verbal testimony; And besides there is the lack of mention of the universal concomitance of fire as the determinant of fire-concomitance (Vahnivyapāyatva). So the condition of Inference is obviously absent in this case. Therefore, the condition upon which the Mimansist has pinned his faith stands invalidated due to its fallibility (Vyabhachāritva).

But the Neo-logician rejoins in a subtler manner –

Suppose – (1) a person observes smoke in the hill

Again – (2) smokehood and sub-concomitanthood appear synchronously in the perceptual cognition.

In the given situation- inference takes place as normal affairs. Now it is evident that the perceptual cognition has to be admitted as the condition of inference.

As the Mimansist is protagonist of 02 cognitive theories in lieu of synthetic judgement as condition of Inference by the logician.

Further the logician posits – if 02 cognitions are regarded as the conditions of Inference, this would do away with the necessity of assuming the element of invariability in connextion of the determination of the causes of Inference. The simplicity which is bound to accrue due to non-assumption of invariability – should be deemed as an adequate ground upon which the Mimansist may repose his confidence.

The logician counter-argues that the utility of synthetic judgment as an essential condition of inference cannot be repudiated on the plea by the Mimansist. However, even the Mimansist cannot evade the responsibility of admitting the element of invariability which forms the cause of Inference. A logical situation based on verbal synthetic judgement has been expounded immediately before to lay bare the futility of his 02 cognitive hypotheses.

Since this type of synthetic judgment which has been occasioned by verbal testimony evidently brings to light that there is the absence of the cognition which has been affirmed to be the condition of Inference by the Mimansist. If he is adamant to exhibit the universal applicability of his 02 cognitive theories he would be compelled to make an unwarranted assumption. Meaning thereby, he would be placed under the obligation to hold that there does take place the cognition which has the determinant of sub-concomitance as its object.

But the logician has made an emphatic denial. He, further, argues that the simplicity of determinant property with reference to causal relation adequately justifies the condition of synthetic judgement as the condition of inference. The cognition holds that the smoke existing in the hill is characterized by 02 distinct aspects viz.

(i) smokehood

(ii) sub-concomitanthood

But the logician should have advocated the notion of universal applicability to vouchsafe the uniformity of the Rule propounded by him. The variability of each and every logical occasion is bound to engender confusion of thought. Now there are cases in which Inference does take place on the basis of 02 distinct cognitions.

If so, how the logician can expect to account for its smooth passage by relying upon his synthetic judgement?

Admittedly the logician cannot cleave to his own fundamental point of view. He, the logician, retorts that the specious argument cannot cut much ice with him (Mimansist).

Here the awareness of complexity operates as an impediment with respect to the idea of causal relationship. Upon the apprehension of causal relationship hinges the apprehension of complexity.

When the apprehension of complexity owes its existence to the awareness of the relation of cause and effect between 02 events, say – (i) synthetic judgment and Inference. So how can this notion of complexity expect to play the role of thwarting agent with reference to its originator or initiator. Now the logician undergoes incentive analysis of the issue in question. He observes that both cause and effect as such are circumscribed by their determinant property (Avachhedaka Dharma). The assertion that 02 given entities are related to each other as cause and effect implies that the determinant property residing in them is strictly delimited to its intended locus. Besides there should be a positive assurance that the determinant property in question is a simple one and as such does not involve any prolixity of thought. Consequently the case of Inference which is regarded by the Mimansist as emerging from 02 cognitions is bound to depend on the synthetic judgement. But the existence of the synthetic judgement has not been testified by a distinct organ of cognition. Now the determinant property of synthetic judgment –qua-cause of Inference is the fact of being a synthetic judgement (paramarshatva). And its locus is the synthetic judgement by itself. It has already been adverted to above that the presence of this synthetic judgement –qua-locus in the Inference taking its rise from 02 cognitions is not attested by an independent cognitive organ.

Summarily the complexity in the present case is intended to be shown with reference to the fact of being a synthetic judgement. And what is its salient feature is that its locus has not been verified by a distinct cognitive instrument in the case in hand. The labour of apprehending it in such a cumbrous way which entails the defect logically termed as complexity of thought.

The inference is a posteriori in its nature as the cause would be inferred by means of its effect-qua-probans.

The synthetic judgment occurs immediately before Inference or the inference take place immediately after synthetic judgement.

Now it is obvious that the present inference is a case of ratiocination from cause to effect. So it ultimately transpires that the awareness of the causal relationship precedes the awareness of complexity.

The former is a fait accompli while the latter makes its presence felt. Consequently the latter ceases to be an impediment to be reckoned with. However the apprehension of causal relationship fails to establish its priority, the consequent determination of complexity is bound to remain unrealized. Here the reason is crystal clear.

The determination of complexity presupposes the awareness of antecedent existence of the synthetic judgement in the case in question. Now the awareness is pendent of the consciousness of causal relationship which has been shown to be absent by means of conditional assertion. So it turns out that the impediment has not been succeeded to present itself in order to counteract the apprehension of causal relationship.

Now Mimansist Prabhakar Mishra has raised vital objection as –

The smoke-qua-probans is not in contact with organ of vision due to remoteness or other intervening factors. He observes that an analysis of the nature of synthetic judgement reveals 02 constituent elements viz.

(1) The knowledge of the probans as existing in the subject

(2) The recollection of the universal concomitance.

So far the probans is taken into account, there does exist ordinary contact of visual organ with it. But the type of contact of the visual organ with the universal concomitance is an extraordinary one.

That extraordinary contact itself is not competent to produce any perceptual cognition. The existence of ordinary contact be dispensed with even in those cases in which the perceptual cognition is about to take place by means of extraordinary contact. If so, how the synthetic judgement can be expected to take its rise when the probans remains beyond the ken of visual organ. This is not even the faintest possibility of the emergence of synthetic judgement when the Inference is intended to be drawn by relying upon the probans with which the contact of visual organ is not within the bound of possibility. This impels the admission of the fact that inference does occur despite the absence of synthetic judgement in the cases characterized by the lack of contact with the probans. But it cannot be gainsaid that such case provides reliable evidence of deviation from the Law of Causation as formulated by logician.

In brief, the synthetic judgement has no legal right to be regarded as the cause of inference.

But the logician suggests that the Mind by receiving the cooperation of 02 cognitions as by the Mimansist-will be productive of synthetic judgement in question, the stand of the logician may be re-enforced by thwarting as logical syllogism.

Atom is a partless substance because it is possessed of eternal dimension due to inherent cause of the germs of knowledge inheres in it.

For example –Ether

An analysis of this syllogistic structure reveals the partless dimensionhood stands for the probandum and its ever abiding dimension occupies the position of the probans.

Here neither the probans nor the probandum can be in ordinary contact with the sense organ due to their supersensuous nature. If so, what is the way of knowing their compresence or co-existence in either?

If Inference is called into question in order to achieve the present objective, this would turn out to be a Loves Labour's Lost. Since the intended Inference would take the shape of inference of eternal dimension in Ether. In as much as in the event of the absence of this inference, the knowledge of the existence of the probans in Ether would be absolutely impossible. In an identical way inferential mechanism is the only potential means to sustain partlessness in Ether.

The net result of these 02 inferences is that they have proved the bare fact of the existence of the probans and the probandum in Ether. But the fact of their co-existence remains beyond these considerations. And this would deter the emergence of knowledge of universal concomitance. Thus ultimately proving fatal to the arousal of inference by itself. But the authenticity of the inferential mechanism on partlessness on the admission of Ether –is an example of an avowed fact so the knowledge of co-existence of the probans and the probandum in Ether is a vital point.

It may be laid bare in the following way-

Consider the 02 previous inferences intended to prove the existence of the probans and the probandum in Ether. Their inferential knowledge is there and as such this would occasion the recollection of the probans and the probandum as existing in Ether as a matter of course.

Consider again there is the absence of knowledge of fallibility of the concomitance (Vyabhichār). In such an unique situation it is reasonable to hold that mind will produce the knowledge of universal concomitance. And admittedly the recollection of the probans and the probandum as existing in Ether will function as an auxiliary factor - thus placing its service at the disposal of the mind.

Now when mind is competent to produce the knowledge of universal concomitance which is judgemental in character. There cannot be valid objection to its producing synthetic judgement in the case of a remote and unconnected probans. So the claim of synthetic judgement to be considered as the essential condition of inference-cannot be challenged by the opponent (Mimansist).

Whereas the possibility of synthetic judgement through mechanism of the mind has been stressed. But the Mimansist counter-argues that the logician's reasoning is based on false Analogy-

Since in the case under review, the knowledge of universal concomitance may be secured by means of Inference. The process may be said down in logical order –

1st place- infer partlessness in the Ether

2nd place – assign the status of logical subject to partless Ether

3rd place - infer eternal dimension in the Ether.

The ultimate upshot is that eternal dimension has been inferred in parless Ether.

As referred to above – the attribute of the logical subject is the determinant property of the fact of subjecthood (Uddesytavachhedaka).

Besides an analysis of the inferential judgement exhibits that the determinant property of subjecthood and the predicate co-exist in the self-same locus (Ether).

It transpires on scrutiny that the co-existence of partlessness with eternal dimension has been clearly ascertained by the present case of Inference. So there is no valid reason to defend upon the mind in order to obtain the knowledge of universal concomitance. When the intended example has been turned out to be an unauthentic one.

How can the proponent (logician) expects to ascertain synthetic judgment in the case of unconnected probans by its taking recourse to?

Further the logician alleges the charge framing tortuous Inference to declare the co-existence of partlessness and eternal dimension. Whereas the Mimansist ought to have grasped the bare fact that mere co-existence cannot be regarded as tantamount to universal concomitance–the element of invariability (naiyatva) is its vital part.

The inference in question has not taken its stock. So Mimansist's confidence in contrived inference is impeachable.

There are other unanswerable objections against it. The Inference under review pretends to prove the co-existence of partlessness and eternal dimension. But co-existence that serves as the adjective datum of inference –partakes the character of relation. And this relation is no other than the relation between the determinant property of subjecthood (parlessness) and the predicate (eternal dimension). And the knowledge of universal concomitance which is essential for the arousal of inference-makes itself felt as can adjective of the probans being the sub-concomitant of the probandum. Thus it is apparent that sub-concomitance (universal concomitance) appears as qualifying adjective of the probans.

As universal concomitance occupies the position of an adjective with respect to the knowledge of universal concomitance. It refers to its appearance in the shape of an adjective with respect to the synthetic judgement also. It stands to reason that when in case of the experience of universal concomitance, it has been left out as an adjective. Here it is but legitimate to expect that its self-same character which appears in the case of its recollection experience and recollection collectively has identity of their contents. That is why the universal concomitance as the content of synthetic judgement presents by itself in the shape of an adjective.

To sum up, the knowledge of the universal concomitance as an adjective is capable to rendering service to Inference.

Now the Inference in question has co-existence-qua-universal concomitance as its object in the shape of relation between determinant property of subjecthood and the predicate. Consequently this inference of partlessness cannot be accorded the status of the knowledge of the universal concomitance as posed by the Mimansist.

But the logician counter-argues that in the case of inferential judgment, there is the cognition of co-existence of determinant property of subjecthood and the predicate in the self-same locus. And it has been stated in the preceding paragraph that this co-existence makes itself felt as a relation but not as an adjective. The underlying reason as to why it is accounted as such may be furnished in the following ways-

(1) The judgement as a class is constituted of 02 elements viz.

(a) The subject – property (substantive) – Visheshya

(b) The predicate property (adjective)- Visheshan

Consider the judgment – the hill holds fire –

Here the hill is the subject of which fire has been predicted

Or

The hill-qua-subject has 02 parts –

(a) The hill as such is the substantive

(b) The property hillhood residing in it deserves to be regarded as an adjective

The predicate (fire) is substantive and the property firehood that dwells in it is to be regarded as its adjective (fire). Now those which are entitled to be considered as an adjective have been set forth with precision.

And those which do not appear as an adjective and yet become the objects of cognition are liable to be regarded as the relation of the substantive and adjective as referred to above.

Take for instance – the position of the hill and fire in the judgement in question. The hill has been affirmed as the substantive and fire as the adjective. It is worthnoting that the idea of substantive and adjective is dependent on existence of a relation between both of them. It is owing to this reason that the conjunction of hill and fire is viewed as a relation. The case is identical with subject to the relation of hillhood and firehood. That is the say, hillhood and firehood cannot be regarded as the adjective of hill and fire respectively without the awareness of a relation between the 02 terms. Consider the case of Inference by this argument –

It is an admitted fact in inferential judgement – the predicate makes its existence felt in the subject. And the subject can be accounted as such if it is circumscribed by its determinant property. The predicate is inferred in the subject which is endowed with a distinctive character of its own.

To put it in other way round- the predicate, say, a fire is intended to be inferred in the locus which is no other than the locus of the distinctive character of the subject. Its upshot is that the co-existence of the distinctive character of the subject and the predicate appears in the shape of a relation in inference.

Now the expression- appears in the shape of relation or appears as a relation –signifies that the awareness of relation takes place independent of another relation. That is to say – there is no possibility of the awareness of the relation of relation in any conceivable way.

But the case is reversible with the subject to another thing- take the case of awareness of an adjective – the adjective as such cannot be known without the knowledge of its relation. But this relation as such does not stand in need of any other relation for its cognition.

Now it is evident that awareness of co-existence of partlessness and eternal dimension appear as a relation in the shape of Inference inferred to above.

If so, there cannot be the slightest warrant to treat the inference as equivalent to the knowledge of universal concomitance. It ultimately turns out that the service of our mind will have to be called requisition to secure the knowledge of universal concomitance of partlessness and eternal dimension. It is the mind which is in a position to contribute to the emergence of synthetic jugdement in the case of unconnected probans by receiving the participation of the 02 cognitions that serve as a pivot revolving. Here the extra-ordinary contact through the mind would lay pretension of producing the knowledge of external cognition (Vishaya). While the service of the unique visual organ is another auxiliary factor to cooperate the mind to produce the synthetic judgement. Here the essential nature of the instrument of knowledge produces valid cognition. But the fact under discussion hinges on an intermediate stage with which the intended product cannot be brought into existence. The intermediate stage between the **instrument-qua-cause and effect-qua-valid cognition** receives the designation of functional role (Vyāpāra).

This functional task is the direct product of the cause and is essential condition of production of the effect for which the cause is responsible. In other way, the functional task being the operation of the cause through which the cause succeeds to produce the effect. So it transpires that the functional role is related to the cause and its effect and to its effect as its cause. An instrument (Kāran) cannot be regarded as such unless it is endowed with functional property.

Example – the cause of visual organ –

It is the instrument of knowledge and as such is accompanied with a functional role of its own. This partakes of the nature of a contact with the object (Sannikarsha).

It is an admitted fact that in the event of the absence of the contact, the visual organ cannot be productive of the Visual cognition. In reversion – if **cognition-qua-extraordinary** contact were intended to be an instrument of knowledge, this would have entailed the admission of its invariable association with functional task. But a close vetting exposes that **cognition-qua-extraordinary** contact does not conform to the condition.

Since there are occasions in which cognition-qua-contact (Upnaya Sannikarsha) produces cognition independent of a functional task. It is only with respect to certain cases in which the admission of functional task becomes inevitable as a matter of course. A concrete instance by Neo-logic critic Jagadish Tarkalankar will certify the authentically of functional task to dispose off the syllogistic premises with relative character of the probans -

(1) **Suppose** – there takes place the recollection of the probans

(2) **Suppose again** – there arises the recollection of the universal concomitance and its existence (probans) in the logical subject.

Here it is an indisputable Law of thought that in the case of 02 termed relation, the knowledge of the one occasions that the other. And this fact affords the raison d'etre of the 2nd.

(3) **Suppose again** – there comes into existence a synthetic judgement which (3rd place) may be regarded as the cognition yielded by the cognition-qua-contact.

Now in the present situation, the intermediate recollection may be treated as functional task (Vyāpārā).

Since it is the product of the recollection of the probans and at the same time is the cause of synthetic judgement which has the recollection of the probans as its cause.

But there are cases of the emergence of the similar synthetic judgement in which the intermediate stage is wanting.

(4) **Suppose again** – there occurs a case of recollection which refers to the probans, the universal concomitance and the existence of the probans in logical subject as its contents without recognizing temporal sequence.

(5) **Suppose again** – the synthetic judgement takes place on the basis of the previous recollection when synthetic judgement has been brought to its existence by means of recollection. There can be no objection to hold that the **cognition-qua-contact** is nothing but the cognition by itself.

In the present situation, the need of intermediate stage has been leftout of consideration. So it is evident that the utility of intermediate stage in the matter of the production of synthetic judgement even in identical type is not a uniform one. Consequently it is clear that the cognition-qua-contact is not invariably connected with functional role in the case of the production of the synthetic judgement. If so, how can the cognition-qua-contact be accorded the status of an independent instrument of knowledge (Pramāna).

The gist of the whole debate is that the **cognition-qua-contact** is susceptible to the mind. So it cannot be treated as an instrument of knowledge. Since it does not conform to the criterion of an instrument (Kārana) counterposed by the Mimansist.

Now the logicians have anticipated another formidable objection – they observe – well, we are prepared to admit that there are cases in which an intermediate stage is conspicuous by its absence. Consequently it is evident that the cognition-qua-contact may not have a function at the time of its producing the synthetic judgement in all cases. But there are cases in which cognition-qua-contact produces a functional or operational task occasioning the synthetic judgement. And this cognition-qua-contact complies with the conditions which are essential to constitute an instrument. So it cannot be denied the rank of being worthy of an instrument to knowledge.

In this peculiar situation, the logicians aver that – if the cognition-qua-contact owns a functional task of its own, it would be subsumed under perception which is an accredited variety of the instrument of knowledge. It deserves to be stressed that if the substantive and adjective of a judgment are purveyed by the cognition-qua-contact, it would be the case of mental perception. It would be improper to treat it as an additional variety of the instrument of knowledge. If the resultant knowledge has specific character (vaijātya) of its own, this would justify to invest its occasioning instrument with distinct and independent status. But in the case under review, there is the absence of its apprehension. So how can it be regarded as something other than the mental perception?

The Mimansist may put forth the plea as –

The mere fact of its non-apprehension cannot have a deep foundation. The mere fact of its apprehension cannot be a ground to deny its existence. A concrete instance may be provided to confirm the authenticity of the assertion. It is testified by empirical evidence that fire can be produced by a dried hay with rubbing to be highly inflammable wood called Arani or with solar gems. Now each and every individual of them were regarded as the cause of fire, this would be at variance with the Law of Causation where the dried hay (Arani) engenders fire, the 02 other recognized generators are absent. The case would be parallel with respect to Arani wood or solar gems.

The net result is that all the remaining avowed cause of fire will be absent during the production of fire by one of them.

The negation of a contemplated cause at the time of production of effect operates as a forfeiture of its right to be accounted for as such. But the verdict of experience has authenticated their claim to be regarded as the cause of fire. If so, what is the way of resolving this apparent checkmate (Gatirodha).

Such checkmate may be solved through the argument below – It is liable to deny that fire is produced by all of them in their individual capacity. It brings to light of the autonomous and heterogeneous character of the causes. So it may be legitimately held that the effect produced by them is also heterogeneous in its nature. And this adequately answers the requirement of the Law of Causation.

To be elaborate, the dried hay is the cause of heterogeneous type of fire which may be termed a hay fire (Tarnavahni). And the same rule holds good with respect to the cases of the 02 other accredited causes of fire. It is true that the observation of fire cannot reveal its heterogeneous trait. But it is admitted of the strength of reason elucidated above. The conclusion is irresistible that what has been evidenced by rational argument, cannot be repudiated on the score of its non-comprehension. So it may be reasonably concluded that there is heterogeneity of the effect produced by the **cognition-qua-contact.** And this again places the claim of **cognition-qua-contact** to be regarded as an independent instrument of knowledge on a stable basis.

Here the heterogeneity of effect seems to be the case of paralogism because there are 02 accredited variety of causal relation –

(a) The cognition of adjective is the cause of the determinate cognition (Vishishta Gyan)

(b) The sense-object-contact is the causal of perceptual cognition.

Further logician seeks to clarify-

Consider the case of recognition (Pratyavigyā) which is a kind of determinate perceptual cognition. It is occasioned by both-ordinary and extraordinary contact. The identity of a thing to which the organ of vision is in actual contact and which was perceived on a previous occasion. Recognition seeks to prove constancy of a thing despite the **spatio-temporal-intervention.**

If 02 mutually class-character co-exist in the self-same locus. This would result in the promiscuity of them.

Now it is an open dispute whether the 02 attributes viz.

(i) Elementality (Bhutattva)

(ii) Dimensionality (Murtattva) – are equally entitled to be accounted as impediment class- characters.

The attribute of elementality exists in 05 varieties of element – earth, water, fire, air and ether. So it is obvious that such attribute may be plausibly treated as universal. But in an identical manner the dimensionality resides in the 1st 04 elements and the Mind and as such conforms to the criteria of class-character.

These 02 attributes are exclusive of one another. Since elementality does not dwell in mind and dimensionality does not exist in ether. But they belong to the 1st 04 elements in an identical manner.

So this turns out to be the case of promiscuity of class-character and as much cannot be accepted as universal. Truly, the rejection of the one as universal evades the chance of the promiscuity of class-character. But in the present case there is the absence of crucial evidence in favour of one and for the exclusion of the other. This is a vital reason as to why promiscuity operates as an impediment to the admission to the rank of universal. It is one of the fundamental postulates of Logic that 02 or more universals are entitled to co-exist in the same substratum only if they are related as genus and species (genre-class).

To be more precise, when one universal is comprehended in the orbit of another, there co-presence does not constitute any logical counter. When 02 universals are related in such a way that one of them co-exists with the other in one substratum and at the same time co-exists with the negation of the other in another substratum, they are related as includent and included (Vyapyavyapka) extensively. Further substancehood (Dravyattva) co-exists with earth-hood (prithvittva) in the self-same earthen pot. The substancehood due to its wider extension not only resides in earthen wares but also in water, air and the like in which there is the negation of earthhood. So it is evident to be co-existent with both – the earthhood and its negation alike.

Now counterargues that the admission of class-character in the case of promiscuity would bring about a breach of the Rule.

Consider the case of the attribution of elementality (Bhutattva) and Dimensionality (Murtattva) –

So far elementatity is taken into account, it does not reside in the Mind in which the existence of Dimensionality is an admitted fact. When the elementality is not existent in the locus of dimensionality, there cannot be the remotest possibility of the former to be the includent of the latter. It is owing to this reason the promiscuity has been held as preventive of emergence of the idea of the class-character.

Counterargues- The validity of the Rule may be tested as follows:-

The ordinary visual perception which occasioned ordinary sense-object-contact, remains beyond the sphere of heterogeneity. But suppose a case of tactual perception occurs by means of both ordinary and extraordinary contacts with the organ of touch. In the given situation, the admission of heterogeneity in the resultant cognition becomes an unavoidable need, it is evident that there is the negation of the attribute of visuality (Chakshuttva) in the perception produced by the sense of touch. This exists that both-heterogeneity and visuality – are mutually exclusive of each other. And both of them are co-existents in recognition. So far the existence of visuality is taken into account, it has been admitted without averse note.

Now if in the presence of heterogeneity were accepted in recognition on the score of its origination from cognition-qua-contact, the co-presence of heterogeneity and visuality would be an obvious fact. And as a result of which there would occur the promiscuity of class-character. And ultimately this would turn to be invincible or unassailable hindrance in the matter of the admission of heterogeneity in the production of the **cognition-qua-contact.**

To round up the present discourse, the **cognition-qua-contact** may be accorded the status of an accessory to the mind in the matter of production of the synthetic judgement by it in the case of unconnected probans.

Neo-logic-critic Raghunath Shiromani proposes to deal with crucial point asserting cognition-qua-contact explicitly. There are cases in which both **ordinary qua-contact** and **cognition-qua-contact** are equally present as the essential condition of the perceptual cognition. Here is it apt to admit as the adjective-qua-contact of the cognition?

Consider –

(1) A piece of sandalwood is fragrant –

It is manifest that there is an ordinary contact of the visual organ with the piece of sandal wood. So far as its fragrance is concerned, it stands in a particular position.

It is an admitted fact that the contact of the perceptual cognition is uniformly in contact with the sense-organ. So in its conformity, it is but legitimate to expect that fragrance also is in contact with the organ of vision. But it is not attested by the verdict of vision. Truly, there are different varieties of contact to account for the perceptual cognition of different varieties of objects. It is universally acknowledged that its conjunction is of a distinct type and analogous being the case with conjunction-cum-inherence. Considered in the light it would not be unreasonable to suppose that there does not exist the **contact-qua-conjunction-cum-inherence** between the visual organ and fragrance. This contact is not in a position to yield the knowledge of fragrance.

Since fragrance as such is not the object of the organ of vision, it is perceivable by the olfactory organ. Here mere existence of contact is immaterial in the face of the fact that fragrance is not amenable to visual perception. The ordinary contact cannot transcend its limitation. That is why cognition-qua-contact has been resorted to as a means to render the visual perception of the fragrance in the judgement.

Further the fragrance has been the contact of the visual cognition in the shape of the adjective. There is no denying that the case in hand attests the existence of ordinary contact also.

That ordinary contact as accredited condition of perceptual cognition has received universal cognition.

And analysis of the whole situation exhibits that both the cognition-qua-contact and the accredited condition of perceptual cognition are equally present in the case in hand. Now when an ordinary contact is clearly wanting in the knowledge of fragrance –the logician holds that synthetic judgement in its entirety is produced by the cognition-qua-contact, provided the probans is not connected with the organ of vision. It is apparent that the ordinary sense-object-contact which is an avowed condition of the perceptual cognition, is an absentee in the case in hand.

(2) Thinkover the case of the perceptual cognition of the colour of pen. In as much as its colour cannot be brought within the ambit of conjunction. So this fact needs the admission of the contact called –

Conjunction-cum-inherence between them.

To be precise, there is conjunction-qua-contact between the visual organ and the pen which inheres its colour.

Therefore the contact of visual organ with the colour of the pen partakes of the form of **conjunction-cum-inherence.**

It is transparent that visual perception of the colour is produced by it.

Even a cursory glance renders it evident that in the 2nd situation the condition –qua-contact lacks any association with the collocation of the **condition –perceptual cognition.**

When the assemblage of conditions of perceptual cognition is wanting, there is hardly any room for dubiety that the expected cognition in question would be a different species of cognition.

To recap the whole situation – cognition-qua-contact is intended to produce a cognition during the absence of ordinary contact. If so, it out to be admitted as the cause of specific type of effect. And its ultimate result would be the claim of cognition-qua-contact to be treated as additional variety of instrument of knowledge. Neo-Logic critic Raghunath Shiromani – he, himself, counter-argues squarely that the opponent (Mimansist) has failed to hit the nail on the end-

The assumption that cognition-qua-contact, if accompanied with the ordinary contact, produces a specific variety of cognition is not based on logical ground or is bereft of point d'appui. In the event of co-presence of both of contacts, the cognition cannot be regarded as the product of one of them to the exclusion of the other. The existence of the cognition-qua-contact and the non-existence of the ordinary one would entail the material difference in the situation. Since is that eventuality all the contacts would come to hinge on the condition-qua-contact in order to make their cognition possible. So it is manifest that there is not the slightest justification to invest the cognition-qua-contact with the status of the cause of a specific type of cognition.

Further the ordinary **sense-object-contact** may be the productive of a perceptual cognition of a thing despite the default of the cognition-qua-contact and vice-versa. It ultimately turns out cognition-qua-contact should not be regarded as an inevitable antecedent of perceptual cognition as class-character.

That is to say,-

Where there is the negation of ordinary contact, the admission of cognition-qua-contact as a distinct type of cause of the specific variety of cognition would be inevitable. Further the causal relationship pertaining to the ordinary contact and perceptual cognition will have to be viewed in a different light –

(a) The object which becomes known by means of ordinary contact is felt in introspection – as I perceive it.

(b) The case is reversible with respect to the cognition-qua-contact

This is not all, But it is expressed by the negation of judgement – I do not perceive.

Not it is apparent that the relation between the cognition with its content in the case of an ordinary contact has its distinctive entity. Its negation in the case of cognition-qua-contact is not liable to dispute.

Therefore both of them cannot be regarded as the cause of the product viz. perceptual cognition in its product of identical character. When the product of one is not identical with that of other, there is no possibility of a breach of the Law of Causation.

Suppose – 'A' & 'B' stand for 02 distinct causes

And 'C' & 'D' stand for their respective effects

Suppose again 'C' is going to be produced by 'A'

It this situatin 'B' is conspicuous or evident by its absence. But this would not entail mismatch or counter or antithetical views.

Since there is also the negation of its effect in an identical manner. To round up the present discourse, the arguments and counter-arguments that have been advanced upto this extent for or against the materiality of relevance of the synthetic judgement calls for a critical evaluation. The logician has succeeded to silence his opponent (Mimansist) by providing logical solution to all his objections:-

He has exposed that the 02 types of cognition theory set forth by the Mimantist stand assailed by the charge of heftiness of thought.

Therefore, the claim of synthetic judgment be regarded as the cause of Inference resting on a solid foundation.

The opponent (Mimansist) reminds the logician that the recollection of universal concomitance must confine itself to the limit imposed by its occasioning experience.

Consequently the experience requires a close scrutiny from the standpoint of the logician. He holds that the knowledge of universal concomitance is required in the wood lit –kitchen.

Truly, there is non compresence of individual smoke and individual fire. Obviously the element of universality that constitutes in its vital breach is wanting there.

The logician has been impelled to resort to an extra-ordinary contact that passes under the name of universal-qua-contact (Samsanya laxan).

The individual smoke, and the individual fire that are observed in the wood lit-kitchen, serves as a locus of their respective class-character.

So it is not liable to deny that the knowledge of smokehood and firehood is there and the knowledge of the 02 universals (universal + extraordinary) will take the place of contact- thus making the knowledge of smoke and fire as a class-character possible.

And as a result of which there occurs the knowledge of universal concomitance of smoke and fire as a class.

The recollection of this universal concomitance as an instrument to acquire the knowledge of an individual smoke being sub-concomitant of fire, this would prove fatal to the position of Inference.

It is attested by experience that what is possessed of smoke is possessed of fire.

Again the service of universal-qua-contact may be called into requisition to yield the knowledge of all things that are possessed of smoke. And the case is identical with respect to fire also.

The impression has been produced in the Mind on cause-effect-relationship. The fact is undeniable that synthetic judgement has universal concomitance as its adjectival reference. And this universal concomitance in its turn, has the fact of holding fire (Vahnimattva) as its positive part.

It transpires on the ultimate analysis and evaluation of the things that synthetic judgment is one of the nature of the knowledge of adjectives due to its reference to the fact of holding fire as its object.

The inevitable consequence would be the knowledge of the probandum (fire) in the hill (logical subject or knowledge-base) being perceptual in its nature.

So it is crystal clear that the Neo-logician's argument to comprehend synthetic judgement has delivered a pre-finishing stroke to the very mechanism of Inference.

In brief, this synthetic judgement is nothing but the brain output of Neo-logician Gangeshopadhyaya who encoded it as –

(lingparāmarsha (Vyāpti-vishishta-pakshadharmta Jnanam, Hetu) – distinguishing or characteristic token or symptom or predicate (smoke indicating fire) in the knowledge – base or logical subject (hill) warranting invariable concomitance. It serves to conclude what is to be adjudged with inferential instruments under 03 conditions of symptom-

1st token /mark – the knowledge of major term in the knowledge-base i.e. smoke in the hill.

2nd token /mark – inseparable connection between smoke and fire Dhumovahni Vyāpya.

3rd token /mark –the knowledge of the pervasive cause as essential connection.-Vahni-vyāpya- dhumvān.

In roundabout way, he admits the propriety of the application of inference next to perception being the experience of cognition caused through long tradition.

He emphasized on Parārthānumāna (inferential process for other's based on invariable concomitance or synthetic judgement or causal symptoms): and the other being Swārthānumān (assumption of one's faith or belief but self-centred).

Natural Intelligence and Artificial Intelligence

Navya-Nyāyology is the study of :

(a) Reasonings, logical propositions and data analyses.

(b) Definition, classifications, divisions, interpretations and co-activities.

(c) Considerations and language programmings.

These intellectual treatments are the gift of the brain and mind as everchanging hypotheses.

Let me treat the Brain Biology being physical to control emotions, reactions, movements regarding the Mind (non-separate). It has at least 05 parts as receptives to think and realize on environs-

(1) Cerebrum – larger part

(2) Cerebellum – in the posterior of the brain below occipital lobes.

(3) Diencephalon – deeply located within the brain connecting the mid brain and the fore brain.

(4) Brainstem – connected with (a) spinal cord (b) mid brain to regular eye movement with visual and auditory information.

(5) Medulla oblongata – the base-part as a control-center for the function of heart and lungs.

Let me distinguish reactions prone to left and right brain-

Left	Right
1. logical	emotional
2. Orderly predominates planning	Focused on Arts and creativity, imagination
3. Calculus Mind preferring to authentic knowledge	Preferring fabricated story telling
4. Well authentical language	Interested in social work
5. Deductive result	perceptive
6. Prone to philosophical facts	Prone to video devices.

Thus brain is informative tools with mind and consciousness furnished.

Further natural human brain has advanced conscientiously prone to survival impulses, mobilising rational concerns and maintaining needs of nature. In Tantrology the stimulation of 07 Chakra-points mechanized for basic focusing inside the brain leads to the explication of the cognitive science to exercise syllogistic arguments.

Such cognitive science may be divided into 02 facets –

(1) Brain Science as -

- (a) Cognitive Convention
- (b) Perception
- (c) Learning
- (d) Memory
- (e) Consciousness

(2) Artificial Intelligence –viewing humanoid contrivance as –

- (a) Learning
- (b) Reasoning
- (c) Perception

Further specific regions for brain intelligence are prone to the most powerful correlation of mass and intelligence frontal – temporal-portal lobes of the brain. The brain science (neuro-science) is the study of the nervous system having 86 billion neurons functioning through perception, memories and emotions being its products.

Whereas Artificial Intelligence (AL-9) machines do not have natural mind and consciousness. But they impersonate intelligence and presumption for the purpose.

At the same notation, the Neuro-Scienc of Intelligence incorporates correlating external causes to intelligence and Brain-volume application within MRI (Magnetic Resonance Intelligence), FMRI (Funcational MRI), EEG (Electro-Encephalography) PET (Position Emission Tomography) and other Micro-dialysis system of the Brian complex.

Such correlative intelligence may be identified as -

(1) **Complete Brain Volume** – It is unprone to analyze brain volume as one-to-one or constituent mediary of human intelligence variances.

(2) **Cellular Substantia Grisia** – Correlaton of Gray Matter to intellectualize and speculate selective discharge of unused synapses to maintain a better Brain Transistor.

(3) **Cortical density** – here density and intelligence vary between group of human beings but not vary in Cross.

(4) **White matter volume (sub cortical)** – connection between Matter Volume and Intelligence at low potential.

(5) **Neural Energy** – Its main purpose is to unheed inferential information having high intelligence than that of persons with low intelligence. This lowly powerful brain is so complex to generate the lightest level of consciousness and mental measures with which we are able to perceive, act, learn and memorise.

These neural processes are explained to be entities with phenomenal images or consciousness. This consciousness is the memory of mind. These memories form base on which the res reactions of Mind work. This consciousness is not merely a positive cache. The mental complex encircling the innermost sphere of consciousness is observed to be beyond finite enumeration.

Further let me discuss about-how does the Brian work and how working of AL-I (Artificial Intelligence) is related to human brain-

Human brain which has been the major incentive for AL-I trials, browses for decades. The search for regenerating cognitive competence (logical potentiality) of the brain in Deep Neural Network (DNN) prevails one of the tricky goals of AL-I. From that angle the intervention of AL-I and Neuro-Science could be observed as one of the most charming sectors of research in ultra Modern Science.

Here the regenerating human cognitive role should work on coupling neural network fabrications that impersonate neural link to warrant those roles in the Brain.

Today AL-I has grown from data-model for problem solving to artificial neural networks – a computational version based on fabrication and function of human-biological neuro-networks. To be more clear, the fabrication of Artificial Neural Networks is caused by human nervous system that trains devices to hit sense of expressions, icons and models.

Also it has accessed to ventral identification to Xerox or match the neural functions of the human brain.

Whereas Artificial brain proxies and emulates recreational roles in Mechatronics as –

(a) Cognitive Neuroscience – to observe how does the human brain work to adjudge decisions and events.

(b) Strong AL-I – human-like behaviour

(c) **Inference (Deep Learning)** – here the data sources are typically systems to capsulate the live-data with mechanism that generates the data. Inference mechanism is self-reliant Algorithm to derive Conclusions in knowledge base (Pakshatasystem). In AL-I, expert system performs such Inference.

(d) **Neuroscience** – (Connectonics fusing the Theory of the Brain)-here human connectome project is working. Neuroscience and AL-I work as bi-directional connection. Motivation for neural-work designs to regenerate core-building blocks of human intelligence. One may apply 04 initial modes of Inference.

(i) Implication basis/ Elimination/Modus ponens as bi-conditional in deductive arguments.

(ii) Conjunction

(iii) Disjunction

(iv) Hypothetical syllogism

At the same time Algorithm – is a systematic logical approach to solve tasks as a set of step-to-step procedures. Its synonyms are - Programming, Inference model or Bayesian Wavelet (probabilistic, stochastic).

These steps are –

(a) Data input

(b) Data processing

(c) Conclusion or judgement output

The above probabilistic theory is a type of reasoning (Tark or Hetu) that considers either the similitude of an event happening or the density of one's faith about an event or premise.

Further computational Neuroscience links the gap between human intelligence and AL-I by generating hypothetic version of the human brain for polymathic reviews on its roles as well as motions, visions, learnings and sensible stimulations.

(6) **Roboethics -** human ethics is applied to Robotics. Beyond neurons other cognitive functions of human brain treat to receive with AL-I as follows:

(a) Consciousness – is one of those supernatural virtues or faculties that we fail to observe well for what our brain mechanism concedes us to cast on a particular job and by-pass the rest biome as a source of motivation in curvaceous as curvilinear neural network as well as Deep Reactive Memoir criterion.

(b) Ephemeral Memory – when we recall personal memoir-events.

(c) Incremental Learning – techniques used to overcome catastrophic oblivion by inter-learning data from multiple tasks during learning due to the weights on neural network through Elastic Weight consolidation (EWC). Since EWC permits Deep RL (Re-enforcement Learning) network to prop up Vector Machine in ML (Machine Learning).

Before demonstrating the functional mechanism of AL-I, I should decipher the recurring terms for better understanding –

(1) Knowledge – is the compilation or passel of expertise and information that a person has acquired through experience. One may appreciate about the subject – reading, learning, researching and memorizing facts. It is amazing and phenomenal; but it gets disappeared as techniques and technologies waver or blink.

It has the potential to be equated to power. Since knowledge is represented in the mind and the mind is stipulated by the cognition of intellect (Buddhi), inspiration (prerana). Subconscious mind, (Avachetan Man-Memory) when the knowledge acquired through perception is determinate (Savikalpaka) viz.

(a) I cognize a pitcher-relation between subject and predicate.

(b) This is Mohan (present) and that was Mohan (past). The cognition of "that Mohan" comes from recollection sub-consciousness). But an Analogy is common to both. The concept of collecting and utilizing, information to support decision-making in some formal or structured way is nothing new. In order to obtain advantage our mental fuss, it is imperative to possess the most update and accurate information regarding among other things, their intentions and capabilities. This Rule applies in every field-be it politics, business, military strategy or criminal intelligence. In addition to this, it is a process that has always been and still is developing continuously.

In response to changes in social, cultural and technological events, the organisation needs a new level of analytical skill.

Further human knowledge may be implicit which exists within human beings. It does resemble implied or informal type of knowledge. It is complicative to express or transmit or exchange of ideas.

(2) Intelligence – is the skill or talent to use knowledge. One may lack in knowledge of a specific subject but may use one's intelligence to help solve issues. Further it is know-how to foster knowledge and apply reasoning to solve riddles clinging to "Brain power". It is psychic versatility or flexibility to deal problems and plight of life. It has innate gaiety or blazing or flaring to evaluate the procurement of learning. By dint of "Intelligence", we may cultivate cognitive exercise through perception, learning, memory, contemplation and consciousness. Here it would be somewhat out of the mark to narrate the oriental claim of "consciousness". The ambit of "consciousness" depends on its definition to the extent of deep computation.

Human visual consciousness (mind) is experienced when the luminescence enters our eyes and it focused into the retina. Rods and Cones (Photoreceptors) which sense the light and transmit electrical impulses via optic nerve to the brain. We do not perceive the Retinal Sensor data directly. The Retinal Image appears to get written to the Memory "Buffer" (Visual Memory) and our visual experience is what we see from the "Buffer". We are buffering things always. In computer science we use "hard-ware Buffer" and software Buffer (Cache) which might hold data waiting to be written to the disk or data waiting to be read by the CPU.

According to "Yogavasishtha" of Monistic Vidantin Maharshi Valmeeki, the Intelligence is other than self-knowledge but what constitutes the mind. Consciousness may be infinite (universal) or finite (Machine limitation) in moving or unmoving creatures in the space and time. Now such consciousness is mechanized to function in ML (Machine Learning).

Artificial Intelligence (AL-I) on oriental parameter

It is the theory of computer system capable of performing tasks or goals that require human intelligence through logical steps. Human features are transformed or translated into Machine inputs. As human features include.

(a) Voice recognition

(b) Visual perception

(c) Decision-making

(d) Translation

(e) Consciousness

(f) Learning new knowledge based on prior knowledge.

AL-I in ML has the same functionality focusing on learning, reasoning and self-awareness.

Like human or natural Intelligence having multiplex network, AL-I, also has complex behaviour or network comprising of 04 analytes.

(1) Thinking humanly

(2) Thinking Neo-logically in DL (deep learning)

(3) Acting humanly

(4) Acting neologically in DL (Deep learning).

Here, DL works on analyzing inferential prediction.

Further AL-I is technically termed a ML in which information is **universal concomitance, (Vyapti)** in codification of Computer Intelligence as well as Infrastructure (Architectonics). The ML is software codes that execute mathematical algorithms. Here AL-I Inference is the neural network rendering or harvesting effective result through –

(a) Inference Engine

(b) Neural network.

Further its infrastructure is based on Deep-Tech-ML-Technology in Deep **Analogy (Upmāana)** of Phytogenetic science, psychology and neuro-science.

In other way, AL-I is the study of rational or logical human Agent having eyes, ears and other organs which act as sensors to make decision as a person through machine-software. It carries out an action with its optimum outcome after considering past and current percepts (Agent's perceptual inputs at a given substance). The AL-I system is composed of the Agent (Channel) and its environs in a given context or knowledge-based (Pakshata). The Agent may be anything that can be viewed as -

(a) Perceiving its environs (Pakshata) through **sensors (Samvedaka).**

(b) Acting upon the environs through **Actuators (Pravartaka).**

So to comprehend the complex of Intelligent Agent, one should be familiar with sensors' Architecture and Agent Programme. Architecture is the machinery the Agents execute on.

It is a device with sensors and Actuators as logical Agents functioning on **Condition-based (Upadhi)** rules -

(a) Human Agent – our 05 sense-organs which act as sensors and leg, mouth and other body parts act as Actuators.

(b) Software Agent – As key strokes, file contents and Network Packets which act a sensors and display on screen, image-files and consigned or shipped network packets act as actuators. The route by which AL-I often fulfils the tasks is undergone Neural Network. Such Network is a process known a DL deep learning) that works with interconnected nodes in a multi-layer structure. The general step-by-step process on 05 Neo-Nyayic-syllogism of Inference includes set of data using Algorithm for goal informations viz-

1. Analytic AL-I – for sentiment analysis, supplier risk assessment. It contributes to data driven decision-making (conclusion).
2. Functional – AL- IoT Internet of Things – as if Navya Nyayic (Neo-Logic) method for conflicting and mutual assimilation of discriminative analysis with critical investigation to specify-valid (conclusion), solution to minimise manual intervention, human error (Viparyaya) and repetitive tasks (Punarāvritti).
3. Interactive AL-I – Chatbos Smart Helpers – it is to automate communications without compromising on interactivities.
4. Text-AL-I (Pakshatā) – Recognizing speech to text + machine translation and content generation capabilities.
5. Visual AL-I - It recognizes, classifies and sorts objects and videos into insights. Such computer-system is applied to visualize the truth of tasks or goals.

Now AL-I refers to the imitation or proxy of human Intelligence in Machines that are programmed to behave and think like humans and similate their actions. This term may be applied to any machine that exhibits features linked with human mind-such as learnings and situations solving.

The main feature of AL-I is its implied task. A sub-set of AL-I is ML (Machine Learning) which refers to the notion or perception that computer programmes can learn automatically and adjust new data or fact without being assisted by human beings.

So the main task of AL-I	Learning Reasoning Perception	Artificial human

out of which reasoning (Hetu) as a software system in the vital tool for processing the use of AL-I and knowledge-based (Pakshata) is put for conclusion or judgement from available knowledge through Deduction (Nigaman Logic) and Induction (Āgaman Logic).

The knowledge (available) may be authentic or accurate which exists outside human beings. It does resemble formal type and obvious to express or transmit or exchange.

Such knowledge is served in AL-I in the way below –

(a) Objects – universal or habitual facts to be true.

(b) Events – occurrence of things in the real world.

(c) Knowledge-base (Pakshatā) – the focus of content of the Agents.

(d) Knowledge-rendition – Positive knowledge to be applied to machines and reversible as advance again.

Further, AL-I is classified as –

(i) Artificial Narrow Intelligence (ANI) – It helps to perform entrusted task freely as humanoid skills.

(ii) Artificial General Intelligence (AGI) – for adults – this system enables to build multiple capacity needed for training by replicating our multifunctional skill.

(iii) Artificial Super Intelligence (ASI) – Exceeds human Intelligence. Here Inference (Anuman) takes place in intelligent Computer-logic from prior knowledge created by evidence to generate new results and deduction.

(iv) Reactive Machines – They imitate our mind's skill to react to multiplicity of inspiritment or animation (Anuprānam). These machine do not have memory-based service rather used for auto-reacting to a finite set or blend of inputs. Such system would be extremely better at every event due to massive memory (Recollection Smriti), faster data-analysis-prone-capacity.

(v) Finite Memory – It enables learning from archival data to cause findings. Probably all existing applications assemble with AL-I. In AL-I system such as those applying DL (Deep learnings) are skilled or adroit by being wide volumes of learnings-data stored in their Memory through Inference Model to tackle problems. Its applications are chatbots (conversational tools for natural language processing through voice commands).

(vi) Mind Theory (Smriti Siddhanta) – It is the next-level of AL-I system enabling to acknowledge the entities or interacts by acknowledging their needs, beliefs, emotions and thoughts process as opposed to Artificial Emotional Intelligence.

(vii) Self-Consciousness (Atma-Chintan). This being the final stage of AL-I evolvent existing hypothetically. Alike human Brain having self-consciousness, AL-I would be able to generate Impressions (Dhāranā) like self-presentation (Atmaprastuti). In ALI, Knowledge Reasoning (KR) is needed to exemplify to predict rationally a priori as human brains behave. It is a software system as a controlling Agent to generate results or conclusion through rational coordination among operations, device control, interface and purpose translation at a given premises or knowledge-base techniques (Pakshata).

Inference Engine & Machine Learning - ML

Neo-Nyayalogy is the base-model of AL-I. It is relevant in today's technology through oral and visual tradition. AL-I can make Inference and draw conclusion without human intervention. Inference that involves inductive reasoning (Agamanatmaka Tarka) to allow programmers and developers to create system for producing logical results in a given situation.

AL-I character is to simulate human Intelligence as a process through computer-system and machines design-software to cognize and behave logically sans human (natural) Intelligence to be more clear. AL-I is the creation of computer programmes similar to that of human beings.

It may recall component terms of Inference suitable to its technical assimilations (Samyojan) –

1. Implication (Nihitārtha) – One of the logical connectives which can be presented as –

   $A \rightarrow B$

2. Converse (Vyatireka) – right hand side proposition goes to the left side or vice versa –

   $B \rightarrow A$

3. Contrapositive (Prati ghanātmaka) – the negation of converse termed as contrapositive –

   $-A \rightarrow -B$

4. Inverse (Vyutkram) – The negation of implication – B/A of A/B. The term Inference is using the information at hand to make logical deduction and prediction.

In other way, it refers to the process of drawing conclusion from data or evidences) (04).

In Inference, we need intelligent-Computer which can create new logic from old one (prior experience) or by evidence (witness – Pramana). Broadly Inference is the conclusion from evidence and facts.

Further Inference rules are applied to derive proofs in AL-I and proofs are sequence of the conclusion that leads to the desired goal.

Let me discuss about 05 organs of Navya-Nyaya philosophy with AL-I simulation in modified fashion –

(1) Learning Agent – Reactive Machine – Machines and computer system memorize specific data or new meterial. Specifically advancement in Deep Machine Learning now enables amplification in prescriptive and predicative analytes through the use of operational data. Machine Learning can find hidden correlation in various data with information and network (05 organs), which can create predicative model (Sādhya) enabling to pinpoint future machine failure.

Machine Learning entails Inference to proceed with running data points to work out an output such as single numerical score-operationalizing a "machine learning model" into production. Since Inference is parametric neo-logic to deduce new and intent factors from existing pre-positions. ML deals conditional rules (Upadhi) already discussed.

Inferential statistics concludes from the sample (Given situation) instead of the entire data. Such learning agent has 04 conceptual units –

(a) **Learning factor** – accountable for advancement from the environs.

(b) **Critic** – It takes feedback from critics to ascertain the accuracy of the Agent or channel regarding deployed standard of performance. Since accuracy probes the gap between the information and reality.

(c) Presentation factor – accountable for preferring external action.

(g) Checksum Detector Generator (Cyclic Redundancy Check – CRC)- it is the process used to detect errors (Viparyaya) in transmitted data. When the Receiver's information mismatches the Sender's data, digital signals suffer from noise that can render errors in the binary Bits. Such factor is accountable to suggest operations leading to new information on causal experience.

(2) Reasoning (Hetu – working Memory capacity) – It uses the ability to make inferences when applying reasoning based on commands of information at its disposal. It serves as the best single predicator of reasoning ability to decipher the systematic variance attested by reasoning or fluid intelligence.

(3) Problem Solving (Theory of Mind) – Used for predicating others thoughts and actions more important that knowledge in solving problems. Theory of mind defines as the capacity to comprehend and consider of other's mind-reading. In advanced applications, problem solving technique in the context of AL-I can include the revolution of efficient Algorithm (Navya-Nyayic) and performing **root-cause analysis (Vijaan-Kuran-Nyaya)** with the task of locating a preferred position.

(4) Perception (Self-Consciousness) – Occurs when real or artificial sense organs scan the environs – Exp-AL-I scans the environs through sense components such a temperature and sensor and cameras. After capturing element of the environs a Receiver will amplify different substances, draw out their functional features and scan **the relationship among them (Samuhaavalamban)**.

(5) Processing language (Natural language processing – NLP)- Machine Learning Technology that renders the computer capability to decipher manipulate and understand a natural language as human beings can. The Assembler takes the task code as input and renders real locatable machine code as output. Human beings have 02 language centres on the left side of the human Brain behind our ears to decode speech.

AL-I processing language is simple as spell check and autocorrect in red or green text (spelling or grammar errors). Language irregularities are scanned through neural network programming. NLP is the combination of AL-I and linguistics to communicate with machines for using natural language. Such Natural Language Understanding - NLU tries to draw out the Sementics (Shābdarth vigyan) of the given text to resolve ambiguity on mathematical proximity as:

(a) Lexical ambiguity

(b) Syntactic ambiguity

(c) Semantic ambiguity

(d) Anaphoric ambiguity

In other way Machine Learning (ML) is the sub-component of AL-I. It may be defined as the Algorithm (Linear Algebra) inferring or parsing data sets through Navya-Nyayic syllogism in a locus and its further deepening to be occasioned from prior knowledge to conclude new decisions. Further in ML the computer programme learns from experience by performing some task (Prior knowledge) and observe how those tasks advance with experience similar to Navya-Nyayic canons. In ML the problems are broken down into several parts and each part gets tested to have combined results. Their linear regressions and decisions are easy to decode and use broadly. In addition to such functionality Deep Learning is also sub-set of ML. DL is focused on gaining more forte by learning how to perform in the stairs of conceptual informations. Further it demonstrates how the concepts are connected with more transparency and how there can be less conceptual renditions for precision. It works on the model of unbroken or stable data-analysis with Neo-Nyayic framework akin to human Brain Processor. It works on many layers of Algorithm called Artificial Naural Network. These network (ANN) and Biological Neural Network (BNN) are alike in function.

Inference Engine (IE Anuman-prakramaka)

Inference is Deep Learning or profound understanding to work out desired conclusion through parsing the data modules. Such is actualized through "Inference Engine (Processors) Anuman prakaramaka that applies Neo-Logic-Axioms to the locus or knowledge base (Pakshatā) to evaluate new data. The machine can apply Neo-Nyayic organs of Inference (Algorithmic Pattern) to detect and designate new representations or images (Inference) on experience and recollection subject to invariable concomitance despite unrevealed or hidden situation. Without Inference the machine would be unable to innovate decision-making skill. Human inference works within cognitive psychology to guarantee the truth of decision over conclusion or findings in the locus.

Further oriental inference is a process of reaching at a valid conclusion by inducing invariable Concomitance (Vyapti) or DNN (Deep Neural Network) composed on multi-layers of interconnected Artificial Neurons (Vyapti). Such DNN (Vyapti) concedes the intent of AL-I task by feeding data-conclusion in a trained DL model being the subject of ML. It is through inference (Anuman) that an image is fed into DNN (Vyapti) through Deep Learning Inference (DLI) to predict against previously unseen data (fire).

Such function corresponds to the organs of Inference as –

(1) The perception of Reason- Major term – smoke in the knowledge-base (Hill-Pakshatā).

(2) **Recollection (Smriti)** – of the reason (Hetu) of Vyapti (DNN) between smoke and fire as experienced earlier.

(3) **Deduction** – Inference of the existence of unseen object (fire) in the logical subject – hill. So here fire is inferred being broadly based on recollection in mindset (Smoke) and the experience of Vyapti (DNN) being the Deep Neurons derived from Memory to avail image with prior perception, notion, apprehension or impression through experience conceived. Further AL-I machines are manipulated to learn from available data information (Smoke in the hill), experience and experts via reasoning framework (Hetu). Here reasoning framework or distinguishing mark or token (linga) or synthetic judgement (Paramarsha) is identified with

recollection (Smriti) of invariable concomitance (Vyapti-DNN) – here we would take heed of the reason that "the term synthetic judgment" devised by oriental Neo logician on the basis of Brain Biology is to get to the truth of judgement derived through 02 parts mainly:

(i) **Right hemisphere** – Referring to complex and holistic-analytic- argumentative synthesis.

(ii) **Left hemisphere** – referring to algebraic physiography as a synthetic genre. Thus algebraic physiography devised the term "synthetic judgement" works as a middle bridge between major and minor terms.

When inferential conduction (Anumān Samcharan), provision, execution or filum generates valid arguments through clues, assumptions, antithesis, reverse or inverse proposition and the unit of the truth or accurate rescuing score is drawn in Navya-Nyayic application.

Further, the Inference Engine – technology is the integral part of the system to fit logical structure to the knowledge base (pakshata). While inferential mark or synthetic judgement qualifies the locus of Inference for new-intended information or knowledge – data. It helps to release information out of the situations or premises and a new information or conclusion takes place. The main duty of the inferential or reasoning Engine is entrusted on 02 models–

(1) Limited Hypothetical syllogism or Conditional Argument Rule – This Inference represents knowledge maneuvers or machination to infer aphorism. This information genre of reasoning entails a limited volume of searching and watching operations in computational domain. Here the output is conditionally concluded.

(2) Proven facts – Proceed with seen or known data and argue for unknown (new) data via Block Chains (Avachhedaka determinant property) or Blocks –prior information for subsequent data being the function of Anumān-Inference in human learning.

(a) **Forward data analysis** – Here Neo-Nyayic reasoning (Tark-Hetu) starts with known data and concludes new data i.e. retroactively posterior or done forward or newer-memory information with retrieval of older memory. Here data-driven block-chain moves from cause to effect (Kārya Kārambhāv Sambandha) or inceptive to decisive, as if, Shinghāvalokan Nyaya (the maxim to follow for advance (new) knowledge (prey) with the retention of preceding (back) or retroactive glance.

(b) **Backward data analysis (proactively) via logical reasoning** (Nyayic Hetu) – Conversely it works in opposite model where the goal is to adjudge the initial data and information already known with the aid of newly known data in given premises. The Engine (data Inference) searches for antecedent (known-smoke) to cause valid conclusion (Fire). This is goal driven to conclude on known facts.

Thus Inference Engine (Anumān prakramaka) is represented through 03 Successive steps –

(a) 1st step (Major term) – Match rules – antecedent (known-smoke) matches data in the knowledge-base (pakshata – already discussed).

(b) 2nd step (Minor term) – Select rule – Inference Engine accomplishes each matched rule in this step and then recapitulate or reiterate back to step – 01 again. This cycling continues until no new (unknown) rules are matched.

(c) 3rd step (Functional rule) – Inference Engine analyzes and assesses the proposition or premises in the knowledge-base to provide goal. Specific task of expert-system involves analysis, diagnosis, designing, scheduling and monitoring to get at the positive answer or goal.

To know how to function deeply – Search Engine in machine learning – as discussed earlier in the chapter "The Definition and Nature of Inference". The cognition of Inference materializes on the nerve of invariable concomitance Vyapti). Where the recurring term-(example) in Navya Nyaya as "Ghat" reveals the physical object – Ghata (Pitcher) sharing 02 character –

(a) state

(b) demeanor.

Further "Ghata" (pitcher) is identified and displayed on the correlation of substratum (Ghata) and super-substratum (Ghatatva) filtered out of billions of substances viz.

(a) Ghata (pitcher or water jar) + Pata (cloth or fibre)

(b) Ghata or Pata

(c) Ghata and not Pata

Details about logical assimilation are dealt in the relevant chapter.

Likewise Search Engine is a big tool to filter intended data on inherent character through unlimited or undesired objects to display the one or ones useful to the users. Here Search Engine operators use logical terms like – AND, OR, NOT and so on.

Here AND – for two or more
OR - for – either of the two
NOT – for – this, not that

Thus Search Engine performs fractions between given time and space through crawling and indexing.

Now it out to be unzipped to aver that inference engine (processor) executes reasoning to enable expert system of the knowledge – base noted below –

(a) Theorem - subject – Major term – Uddeshya.

(b) Hypothesis or proverb - predicate – minor term – Vidheya.

(c) Model classifiers-conclusion – confirmative judgement – Nigaman

Thus, data or statistical or model Inference mechanizes specific skill of knowledge from a interdisciplinary DNM (Deep Neural Model) and infers the goal or result. Here raw (known) data are inputted through a wise Neural Network Model (NNM) and it outputs a prediction (useful information) based on predicative precision of the neural network.

The Navya-Nyaya Technical Language (NNTL) is the product of human computing structured scientifically in Natural Language.

The Integrative Theory of Neo-logic-language encapsulate (1) Akānkshā (Expectancy or significancy), (2) Yogyatā (Compatibility), (3) Aasakti (Application or input) and (4) Taatparya (Purport) as dealt in "Tattvachintāmani".

NNTL is the Semantic Science consisting of 02 features –

(a) the faculty of mechanism of AL-I

(b) the faculty of formulation of AL-I.

Its methodology of analyzing and customizing Semantics is conducive to NLP (Natural Language Programme) and Machine rendition. Its algebraic mode of expression ensures its thematic reliability (Aapta Vachan). Navya Nyaya technical language has been structured for cognitive counter-argument based on logical reasoning. It does treat Artificial language (AL-L) universally. Its focus-centres on pinpoint and assembling-design to formulate an Artificial Verbal judgement is to censor possible ambiguity of Natural Language. Mind that it has been innovated and structured in Depth Analysis for programming with Natural Language by Oriental Neo-Logician Gangeshopadhyaya. The above NNTL's attributive compositions (04) correspond to 04 Computational abstractions in –

(a) Morphemes

(b) Graphemes

(c) Lexemes

(d) Syntax structure.

Such application of Intelligence-based data-abstraction through determinant locks contrives to cognize semantics in a particular frame of reference. Let me illustrate in facsimile:

"Samyogen Sambanddhābhāvachhinna Ghatattva-

Nishtha Pratiyogitā Nirupakah Abhāvān Ghatah"

instead of natural skt. language:

"Samyogen Ghatattvam Nāsti."

The mechanism of its illustrative elucidation has been dealt already in relevant chapter (as original) at suitable point.

On exhaustive exercise of cognitive cudgel I aver the fact that NNTL is a highly efficient device to develop IT sector for programming language. NNTL proceeds with wordkraft sophistication to rephrase for customization at length.

Debate-Argument and Bar + Bench – Intelligence

Shastrārtha (cross statement or argument) in ancient India is called Vādvidyā of 03 kinds-

(1) Vāda – between proponent and his tutor or some others in similar status.

(2) Vāda is the theory of thesis in sound debate where the proponent proposes arguments to prop up his attitude known as Vadina. And the opponent who exposes that attitude through his counter-argument known as prativadina. An approachable or affable-debate materializes when the opponent is wise, benign and conversant with the skill of conviction a dissuasion evidently. Such system is other than dialogues and communications. Here both the parties should exercise their applicable mind to travel over varied scope of the topic in hand through valid rules of the science of syllogism.

The Vāda has the following features –

(a) the premises of the proposition and rebuttal of counter-proposition should rely on an opposite evidence or modes of factual knowledge or precise hypothesis.

(b) deduction or argument should not be conflicting with any standard canon. The both sides should undertake the thesis to be defended and counter-thesis to be contradicted.

(2) **Jalpa – (Verbal altercation or altercative debate or carping criticism or wrangling debate)** – It relates to nitpicking trivial objections, lawful rejoinders and any type of acing for advantage.

(3) **Vitanda (Fallacious argument in controversy)** – a debate where-in no-counter-thesis is put up and the debater contrives to warrant decree by contradicting the proposition posed by the opponent.

Here the propositional canon or conclusion should be based on reasoning, illustration, perception, Inference and evidence to infer by deterring the proposition as make-shifting, relocating, contradicting, rejecting, doubting, irrationalizing or invalidating.

As a matter of note a debater should have the following attributes –

(a) discernment or discretion

(b) Compiling of points of debate

(c) Logical motion of argument

(d) Conclusiveness

(e) Transparency of the intent of the debate

He should be truthful, gentle, cheerful, sedate and prudent.

To be more elaborate – The debate has been analyzed in 16 doorsteps-

(1) The sources of valid knowledge.

(2) The device of precise knowledge.

(3) Fabrication of dubiety

(4) Intention

(5) Illustrations

(6) Stable tenet

(7) Component of syllogism

(8) Reasoned justification.

(9) Deduction of the thesis.

(10) Argument to defend to transpire the facticity or reliability.

(11) Squabbling to warrant decree.

(12) Fault-finding to castigate

(13) Fallacy of reasoning or arbitral antithesis.

(14) Intentional interpretation of the terms to hoodwink

(15) Pointless protect grounded in Analogy.

(16) A flaw in an argument by which a disputant is put down in argument further causing quasi-thwarting.

According to Oriental conception, let me narrate some faults of debate as follows –

(1) Gurvakshara sambandhan – Logorrhoea or loquacity

(2) Parānmukha – Convolution or winding motion.

(3) Trivargena Viruddham – protest of triad-basic aims of human life–

(a) Conducive ethics

(b) Undue life resources

(c) Sensuous pleasure precisely.

(4) Assamkritam – Vulgarity

(5) Nyunam – Too brief

(6) Kashta-shabdam – Pinching usages in forced tone.
(7) Vyutkramā-vihitam – Transgresseive timbre (tone).
(8) Shesham – half done or unfinished.
(9) Akalpena – dubious or cryptic
(10) Nishkāran – Illogical or unreasonable.

It is worthy of note that Buddhist philosopher Vasubandhu, young brother of Asanga, wrote 03 works on Debate Theory as –

(i) Vāda hridaya (core of debate).
(ii) Vāda Vidhān (canon of debate)
(iii) Vāda vidhi (regulatory mode of debate)

Further Debate or Shastrārtha viz. intero-intra debate by great jurist viz. Manu, Yagyavalkya, Vishnu, Kātyāyan, Brihaspati, Nārada, Shukra and Vedantin Commentator Vachaspati Mishra and the like formulated elaborate judicial system and case-trial procedure prevailing since ancient India. They also, prescribed dress codes for the Bar and the Bench in a modest attire being unafraid for the client. The Bar and the Bench should have fidel look, as if , in Yagya hall meant for justice through judicial investigation. Further they used to adopt neutralism to evade attachment, greed, fear, enmity and exclusively hearing in the single bench (lower court) and double bench (3, 5, 7 judges) for valid verdict.

The Oriental works on Jurisprudence concerning Smriti, Shruti, Puranas and Mimansāhastra in the domain of civil law, common law, customary law, law on ethics and hybrid system help to solve social problems in legal terms and values. In oriental legal system, the key-claims of law are discussed in Anthropology.

Smrities focus on the need for adroit judicial system based on righteous (dhārmic) justice. Here Dharma in Jurisprudence refers to righteous duty, social subsistence, legal environs and spiritual fervour in holiness and harmony.

Further the philosophical school of jurisprudence is that of the law classified into Analytical, sociological and theories systems –

The justice system has 04 linchpins –

(i) Dharma (based on truth)
(ii) Evidence (Based on witness or testimony)
(iii) Conventional customs (agreeable to the public)
(iv) Sovereign mandate (as postulated or proclaimed).

The virtues and lucency in the judicial system were exclusive to comply with the rules and regulations. There was the jury system as its salient feature to counsel the court.

The nature of trial was inquisitorial to investigate and interrogate the facts to reach the truth. Under such system the pretrial hearing to bring possible indictment was usually under the supervision of the judge through cross-questioning the parties. The judges were expected to proceed to listen to the arguments before delivering judgment. Whereas adversarial system follows competitive procedure between prosecution and defence to ascertain the facts. It is the natural system to deliver judgement on impartial or unbiased manners with some exceptions. When the witness gets prejudiced the court will judge the evidence under the Rule – "Nemo in propria Causa Testis Esse Debet."

Now Oriental Jurist Kautilya and Vishnu urged the management to attach and seize the entire illegal assets acquired through injustice and betrayal of trust in the public domain. They all were watchful to maintain poise between the science of material welfare and the science of righteousness or rectitude. Further oriental jurists Nārada and Brihaspati formulated the conventional conclusion on dual conflictions cross overlapping provisions as social conditions permit for the purpose.

Because justice means – the exercise of judicial command in exculpation or vindication of rights by determining acquittal or penalty. Justitia (Latin) means righteousness and equity. Shruti, smriti and prescribed uses are the sources of the Hindu Laws. Where 03 types of judgement are classified

(a) **Personam (**plaintiff's role against defendant).

(b) **In rem** (law suit against or about a thing).

(c) **quasi in rem** (as if, against a thing as civil suit particularly).

Here penalty or punishment may be imposed in the following facets –

fines

Confiscation of property

Exile

Corporeal

Mutation

branding

Fortune-loss

Forced labour or

Capital one.

Such types of judgement are formulated in "Charaka Samhita" through the trial of –

(a) Aaptopadesha pramān (Authoritative testimony)

(b) Pratyaksha (perceptual testimony).

(c) Anumān pramāna (inferential evidence).

(d) Upman yukti (analogical reasoning)

These pramānas are helpful to determine known and unknown phenomena in the universe.

Out of them, the inferential evidence is the basic tool of trial and scrutiny to its wide devices through diagnosis, analysis, rationale and interpretation of the legal facts. Further the inferential evidence is employed to Interpret or analyze micro-phenomena to conclude from the known to the unknown speculations in given event. The relevancy and applicability of the process of inference is obliged to the Law of Causation irrespective of the space and time – where cause is unconditional and invariable antecedent of the effect and an effect is unconditional and invariable consequent of a cause. The cause not present in any hidden form whatsoever in its effect. So the cause must be antecedent purvavritti), invariable (Nitya purvavritti) and unconditional (Anyathāsiddhi).

Argument (Parārrthānuman – Devoted to Argument)

Argument is the combo of assertion or premises prone to foster re-enforcement of evidence in one's favour. Here it is in the fitness of thing to distinguish between emotion and argument for distinctive use-

An argument is understood as cognitive and apathetic trial. It is probably based on the dismissal of erring discrimination between argumentative advocacy and emotional reaction. It is an interpersonal presentation graphically, expressively and legally.

I mean Argument should be set up on –

(a) Circumstantial routs

(b) Bi-understanding of the issue in question on basic probe.

(c) Fabrication of claims to be pleaded.

(d) Compilation of evidence in DL (Deep Learning) on multi-prospectives.

(e) Review-arguments in counter-filing and its legality.

(f) Expertise in arguments to convince for ends of hearing.

(g) Paving new logical tricks in argument.

Its validity or invalidity should be adjudged on intrinsic or extrinsic evidences being the successive chain to the valid goal or conclusion.

The Purpose of Argument

It is a device to exchange reasons for one's deduction or inference through legitimate version and/or conviction. It intends to assess the level of the real event or fitness of competitive clarification for deductive end. It is normally formulated typically in causal language beyond natural one. It serves as a group of premises of which the one is asserted to pursue others through valid inference reasonably. The logical mindset on argument is apposite or useful for scientific domain such as mathematical, electro-mechanics and etc.

Remarkable to not here that informal argument is more or less implicit. Whereas the conclusion through formal argument is explicit. It may be seen that the argument is assessed by the audience also in addition to 02 parties.

Steps of Argument

There are 05 steps-

(a) Decompose common proposition into smaller bits

(b) Counter-argument.

(c) Deduction

(d) Elucidation of pitfall reversely (pros & cons) with their reactions.

(e) Reasoning regarding critical analysis through proposal, narration, countering and judgment.

Here mind that qualifier, support and etc. are essential for all arguments. Where model qualifier specifies soundness of the skipping from the data to the warrant and may restrict- how invariably the assertion or argument is solicited.

In oriental logic, prognosis of counter-arguments through argumentative debate with the system of syllogistic Inference auto-trials the counter-moves of the disputant against for each move by the exponent.

Whereas valid argument determines the credible or genuine worth of statistical communication. Since an argument is a chain of communication and the deduction or conclusion is the last one and all its antecedent declaration are hypotheses.

As a matter of heed, the Nyāyology is the science of legal proceedings and judicial sentences through valid arguments as follows:

(1) Pramāna – sources of valid knowledge through –
   (a) experimental
   (b) memory

(2) Prameya - object of provable knowledge in the following:
   (a) Identical – consciousness unit
   (b) Shareer – material body
   (c) Indriya – (sense organs).
   (d) Artha – intended for senses.
   (e) Buddhi – cognition
   (f) Pravritti – propensity, perseverance or validity of rule.
   (g) Man – the mind
   (h) Dosha – mental faults
   (i) Pratyabhāva – afraid of being afflicted in next life.

(3) Sanshaya – (Suspicion or indecision) – Confused state of mind or mind blinking among conflicted views.

(4) Prayogan – (Motive, need or interests) – one acts for either to receive intended objects or act away with unintended objects.

(5) Drishtānta (Case-Citation) – can be admitted by both parties in a debating conflicts.

(6) Siddhānta (True logical conclusion of an argument) – it covers the entire views of logical systems.

(7) Avayava (penta-component parts or syllogism) – They are applied in inferential judgement to have common conclusion.

(8) Tark (vyāghāta) – (Reductio ad absurdum) – it is the process of cross-observation pleading for valid conclusion.

(9) Nirnaya (Verdict or settlement) – to conclude the case by virtue of perception, inference, analogy and discretion.

(10) Bādhit (fallacious refutation) – hybrid decision between propositional and oppositional sides or plaintiff and defendant for healthy conclusion on valid reasoning.

(11) Jalpa (hostile counter or skirmish) – vanity or self-conceit is involved to make convinced the other on his own twist of circumlocution with all hatched arguments in his favour.

(12) Vitandā (Captious or fallible argument) – parties apply such irrational charges on each other in his weak position for chaotic end.

(13) Hetwābhāsa (fallacy or pretention of reasoning) – the reasoning pretends to be valid but not based on validity.

(14) Chhala (Tricks or knavery) – exercising the idea of rufuting the clues of the opponents with his own tricks. The parties misinterpret the bone of contention knowingly to befool others in an implied sense.

(15) Jāti (futile rejoinder) – to show or perform defective augument by the party in question or common or uncommon character in the absence of invariable concomitance or Analogy –

- Ghata (earthen pitcher or water jar) is eternal or non-eternal
- it is produced by clay.
- Whatever is produced is not eternal.
- So the pitcher is non-eternal.

Here Jati has 23 facets-

(i) Vichar Sabhā (trial court or council) where jurors act as referees to confer judgement on criminal trial as per inferential argument.

(ii) Saadharmya Sabhā (Common trial).

(iii) Vaidharmya sabhā (Special trial).

(iv) Utkarsha sabhā (higher court trial).

(v) Upkarsha sabhā (lower court trial).

(vi) Varnya sabhā (later on court trial).

(vii) Vikalpa sabhā (Optional court trial).

(viii) Sādhya sabhā (court trial on Major term).

(ix) Prapti sabhā (findings court trial).

(x) Aprapti sabhā (non-finding court trial).

(xi) Prasanga sabhā (Court trial on connective argument).

(xii) Pratidrishtānta sabhā (crystal court trial).

(xiii) Anupapatti sabhā (Court trial on inapplicable reasoning).

(xiv) Sanshaya sabhā (Irresolutional court trial).

(xv) Prakaran sabhā (Representative court trial).

(xvi) Ahetu sabhā (partially court trial).

(xvii) Arthāpatti sabhā (illogical inferential court trial or circumlocutory).

(xviii) Avishesha sabhā (indiscriminative court trial).

(xix) Uplabdhi sabhā (Court trial on conjecture).

(xx) Anuplabdhi sabhā (non-perceptive court trial).

(xxi) Nitya sabhā (Regular court trial).

(xxii) Anitya sabhā (Irregular court trial).

(xxiii) Kārya sabhā (Causo-operative court trial).

(16) Nigrahasthana (the basis of condemnation) – When the disputant misunderstands his own opponent's propositions and eventually concedes his defeat in helplessness. Now in order to evade the basis of condemnation or defeat, the party should be aware of at least 22 fronts –

1. Pratigyāhāni-Betrayal of proposition to be proved.
2. Pratigyāntara- Deviation from proposition.
3. Pratigyāsanyāsa – Disuse of original proposition (abandonment)
4. Pratigyāvirodha (contradiction of proposition).
5. Hetwantara (Deviation from logical argument).
6. Arthāntara (Difference of connotation).
7. Nirarthaka (unreasonability) it is quintessential element of modern judicial system in legal issues.
8. Avigyātārtha (Bluffing) – Innocence or nescience
9. Aparthaka (Incoherent or senseless argument)
10. Aprāptakāla (Unfit or unsuitable moment).
11. Nyuna (Deficiency of evidence).
12. Adhika (Exaggeration or hyperbole of statement).
13. Punarukti (Superfluity or redundancy of statement).
14. Ananubhashan (non-repetition of an assertion to refute).
15. Agyān (Unawareness).
16. Apratibhā (not ready witted).
17. Vikshepa (Refutation of an argument in his favour).
18. Matānugya (Doubtful excusing)
19. Paryānuyogapekshan (An inquiry with the object of contradiction or refuting the statement).
20. Niranuyogānuyoga (Irregularly non-inquiry).
21. Apsiddhānta (Erroneous conclusion or judgement).
22. Hetvābhāsa (Pretension of reasoning)

In Nyayology an argument has been amplified into the following versions:

(1) Aatmāshratya (self-dependence) – it is an argument which exposes the anomaly indulged into ratiocination which pursues to evince the fact that each substance is dependent on itself regarding its origin or persistence.

Example if 'X' is the cause of 'Y'

if must be different from itself because the cause differs from the effect.

(2) Anyonyāshraya (reciprocal relation of cause-effect) – here an argument exposes the dichotomy indulged in the conclusion 02 substances are mutually dependent on each other viz.

if 'X' depend on 'Y' and

'Y' depends on 'X'.

and 'X' cannot depend on 'Y'. So to say that –

'Y' depends on 'X' – is to reject practically that 'X' depends on 'Y'

Although such notion of mutual dependence has been regained by the old oriental logicians as self-puzzling and unsound.

(3) Chakraka (Circular) – it lies in revealing the fallacy of reasoning where an object is made to elucidate the presumption or hypothesis or pre-supposition – (Tadpekhyāpekhyapekhitva) viz.

if 'X' is pre-supposed in 'Y' and

'Y' is pre-supposed in 'Z' then

to explain 'X' by 'Z' to reason in a circle.

Because 'Z' by it innate limitation resorting to 'X'.

Here the coil or twist of version or demonstration forms a conclusive circle as the cognition comes back to its 1st move through 02 or more intervenors.

Now if I concede that perception is pre-supposed in Inference and the latter is pre-supposed in corroboration or evidence, here to certify perception by evidence is to cause the fallacy of circular reasoning.

In precision – if perception depends on testimony, it must depend on sense-organ contact.

(4) Anavasthā (Absence of finality or conclusion or endless series of cause and effect) – here an argument exposes the idiocy of the messy escalation of the fallacy of infinite presumption. There is an endless ebb or eva of consideration (thought) without any

final locus. Here our consideration moves upstairs, as it were. Therefore, if I render –

'X' by 'Y'

'Y' by 'Z'

'Z' by 'Y' infinitum

I cannot rationalize anything and if I try to deduce the premises of Inference.

I would be entrapped into the fallacy of infinite ebb viz.

If an inference depends on inference for its premises no inference would take place.

(5) Pramānbhadhitārtha (protest of contradictory conclusion or evidence) –

Here the argument ascertains the legitimacy of reasonings that the contradictory of its conclusion or deduction is illogical. Here to protest the contradictory of the conclusion to some facts or common laws.

Therefore –

If its contradictory be fallacious, the authentic conclusion must be valid on a valid reasoning – viz. take in inference case – whatever is smoky is fiery. The substance is smoky. Therefore, the object is fiery.

If this conclusion be false, its contradictory – this substance is not fiery – should be valid.

But the latter proposition is observed to be invalid by the argument below –

If in the case of this substance – Smoke is not related to fire. Then it may not be the effect of fire. There is not 3rd alternative here.

Therefore, it is not due to fire-it must be either self-existent effect or non-existent occurrence.

(a) Here the 1st alternative counters the common law of causation and is therefore, illogical or invalid.

(b) The 2nd alternative suits self-contradictory, since it commits to the proposition that the smoky substance is smokeless viz.

if 'X' (smoke) be a token of 'Y' (not fire) and

'Y' (not fire) were a token of 'Z' (not smoke)

Then 'X' smoke) would be a token of 'Z' (not smoke).

The above idiocy is involved in the rebuttal of the authentic conclusion.

I must deny it as invalid and admit the authentic conclusion as valid based on valid inference.

Here I observe that it is the conflicting type of argument emerging out of deduction of a given thought or its contradiction. So an argument may be applied to look into the cogency of any reasoning – inferential or otherwise out of a given deduction or its rebuttal with or without the authentic propositions.

Lastly the classification of an argument certainly be based on the logical character involved in different cases. Such divisions may be invalid.

According to oriental logician Udyotakar being the author of "Nyayavārtika" suggests 03 types of argument –

(1) Purna sakārātmaka (total positive) – where there are genuine cases to the truth without major term showing non-presence of the reason viz. whatever exists is transient cloud, storm.

(2) Purna Nakārātmaka (Total negative) – showing non-presence of the reason hypothesizing wholly within the logical locus. Here the pervasion or invariable concomitance is contradicted – Exp-

a living body has a soul-heart because it breaches

whatever does not have a soul does not breathe like a pitcher.

(3) Sakaratmaka Nakārātmaka (Positive-negative) – There is fire on the hill since there is waving smoke on the hill.

or where there is smoke there is fire like wood- lit kitchen and unlike pool.

Besides Mahayana Buddhist logician Siddha Nagārguna (Chemist) postulates 03 types of cosmological or teleological argument meant for cause argument and argument from contingency as follows:

(a) Mulmādhyamic Kārikā – Basic wisdom or the mid-way or centrism – Nihilism – Anirvachaniya –

Permanent existing or non-existing realm of dream like appearance or phenomenal and a realm of Noumenon (Absolute reality) – here the object itself is inaccessible to experience to which a phenomenon is referred to for the base or cause of its sense contact. A middle way position between spiritual views that substances exist in some ultimate sense (svabhāva) and the view that the things do not exist at all. The argument is that 1st (intrinsic) leads to the conclusion that ultimately the real things are eternal while the 2nd (non-existent) leads to the conclusion that ultimately nothing whatsoever exists.

(b) Vigrahvyāvartini (Dispeller of disputes) – is said as – I have no thesis for his reaction of positive-epistemic Mādhyamika statement.

(c) Vaidyaka Prakaran (Atomic comminution or Trituration) – The Buddhist philosopher argues in Metaphysics that there is no self-caused entity and that everything dependently arises from or upon something else. All objects and human beings whether in heavenly or hellish or earthly realm are always changing to undergo rebirth and redeath. The impermanence is the source of Dhukkha (distress). The universe is an endless cycle. He asserted a direct connection between dependent origin and non-self/Nihilism. His successor Buddhapalit argued for the use of properly logical syllogism to plead positively for rare voidness (instead of just refuting the theories of others). While Chinese logician Aryadeo opines 02 kinds of arguments –

  (i) ādharmya or Sārupya (Similitude).

  (ii) Vaidharmya or Vairupya (contradiction or antithesis) on arguing blueprint or proposition.

There some key-concepts to model arguments practically –

(a) Carry on your argument in brief.

(b) Ignore crafty strategy.

(c) Mind against grabbling at your argument.

(d) Distinguish between the truth and the presumptive premises.

(e) Prefer logical argument towards the judgment.

(f) Arrange logical evidence to persuade to your favour.

The Experimental genre of argument in involving trends

(1) Induction – it lies on the statistical recurrence which should be universal on opposite inductive viz- an assertion that uses distinctive proposition or observation to form a wide generalization where the conclusion drawn from known premises is valid. Here the past regularities carry on conclusion in future to be true as posterior coherence logically.

(2) Abduction – on considering a few coherent facts, conclusion is drawn to decipher the occurrence of these facts. Although an abduction is deemed to be universal both in science and daily life along with distinct domains like the clinical diagnosis, judicial matter and AL-I. It is based on human cognitive inclination which may be casual to logical complications.

(3) Deduction – the valid premises warrants valid conclusion.

(4) Analogy – comparative argument that interprets the substance in terms of common denominator to others.

An aforesaid explication of argument entails reflection on the science of logic which is a kind of reflecto-presumptive argument to trial the legitimacy of the deduction of any decision. If there is any confliction in admitting the decision as valid, it may be illogical or fallacious or nugatory. On the contrary if I quash it as fallacious the deduction must be legitimate.

The mode of cogitation in logic takes in the conclusion of the illogical premises from a particular locus. It holds the reasoning upshot to lay bare the fallacy of that locus and thereby standing behind the counter-locus.

In logic the inference of fire from the observation of smoke, there may be 02 options –

(1) The smoky substance is flaming.

(2) It is not flaming.

From the latter locus, I infer the premises that the substance is not smoky to be countered by my first hand experience. Such event is demonstrated in the character of hypothetical premises viz.

if the substance is lacking fire.

there must be lacking smoke

Consequently the science of logic corroborates the Inference of fire through the deduction of an invalid premises out of the conflicting hypothesis.

It is the common code that pervader-cum-pervaded relation is based on the attributive or characteristic token / mark. Now in the negation of fire in the substance, I fail to counter the deduction that there is the negation of smoke in the locus that is smokeless. Such a deduction is conflicted by direct perception.

Therefore, it is due to the science of logic defined as the mode of concluding from a distinct token to which it is a token to be untrue.

Now whenever the premises vested by any means of valid knowledge is conflicted or doubted, I should appeal to the science of logic to expose doubt or dissolve the problem on trial.

In argument, I draw the antithesis of the premises characterized by a hypothesis and immediately notice the effect of hypothesis causing antithesis. Because the fallacy of a locus debars the legality of its antithesis.

Further the argument helps to the sources of knowledge but not the sources of knowledge itself.

In this way I see a desk and say –

There is no dish on the desk.

Here, if I doubt the validity of this premises.

I may practically omit it by an argument as if there were any dish, it would have been seen like the desk. But the event is different as non-existence of the dish on the desk due to perceptual knowledge.

Likewise – if the substance is without fire, it must be without smoke. Here the Inference causes the smoky substance to be fiery by the perception of smoke being the characteristic token of fire. Here the hypothetical argument causes such inference. So an argument validates or ensures the testimony of knowledge.

In arguments, the logicians take recourse to scientific probe and licit course of action pervading in many sections of the society. The argument involves into interdisciplinary affairs of investigation.

According to Neo-logician Gangeshopadhyaya interprets- an argument may be contemplated as an independent existence in human-communicative operations. It involves the review and reasoning leading to the process for deduction based on speculative consideration of the given informatics. On the contrary reasoning (Hetu) entails cognitive or psychogenic compass of conceiving the data.

In other words, the reasoning is the bridgeable power between major and minor term to draw inference toward conclusion on circumstances or evidences. While evaluating reasoning for argument maybe classified as below –

(a) Deductive reasoning (from rule to example) – top to down – stepping through theory of hypothesis models- conclusion or preposition, validity, solid argument and conclusion warranting the veracity (Yathārthatā).

(b) Inductive reasoning (Bottom to upward) it goes through the series of certain data and concludes on cause-cum-effect reasoning.

(c) Adductive reasoning – here the propositions conclude on hypothesis of various truths facts, texts, scriptures and literature conjointly. Scientists and logicians apply such reasoning to design computer models in AL-I. Such reasoning may not be true but helps to search about multi-hypothesis to form precise purview known as –

(1) Savyabhichāraka Graha (fallacious instrument or violative norms) – Biped human beings are rational

pigeons are biped.

So pigeons are rational.

Here the common factor-rationality falls in question.

(2) Upādhi (adjectival character or discriminative property) – Because the wet-wood pieces (adjectival character) can emit smoking for fire and not effective in general (dried wood).

So here special kind of cause occurs in causal relation of smoke and fire in special condition.

Further, the argument pertains to the common compass of the trading of reasons intended for the assemblage. Although neo-logician Gangeshopadhyay evolved the thought of Vedantine Vachaspati Mishra in the following facets –

(1) Dynamic legal argument –

Syllogism

Major term – general rule

Minor term – claim about specific case

Conclusion – application of general rule to a particular case.

(2) Oral argument – debate on spoken presentation.

(3) Reasonable thought – to adjudge for legal reason

The Fallacy of Arguments

At least 04 types of fallacious arguments broadly –

(a) Fallacy of amphiboly (Ubhaychara) – Here a debater takes advantage or manipulates the pun of alleged terms to be treated favourable to one's side out of their anagram in order to draw superfluous conclusion.

(b) Beating around the bush or assuming the truth of conclusion rather than proving the same or begging the question –

Where one of the propositions and the conclusion being the same but contrarily or otherwise worked out or hatched up.

(c) Pleading to the controlling authority – here the averment is seconded with reference to the authority instead of posing reasons in its help.

(d) Ad hominem (irrelevantly attack of some aspect of the opposite party while arguing instead of advocating one's position) – It involves cynical side of debater to argue against the opposite advocacy.

(e) Faulty analogy (Similar mistakes) Two things are alike in other ways just because they are alike in one way. Here the fallacy is not in the structure of inference rather in something else as brother-sister-may have physical similarity with some differences otherwise. There lies not ample likeness between the derived sphere and the game sector.

In other way, the argument may be fallacious as follows:

(a) Anaikantika (Skeptic or aberrent) – the Neo-logician explains it as savyabhichar i.e. too general middle term for explanation as multiplicity of objects. It may be –

  (i) Common (Sadharan) – Fire exists in red hot iron bar which has no smoke. Here the pervasion is debarred by the invalidity of the common.

  (ii) Uncommon (Asadharan) – The invalidity of the uncommon exist in the reason which fails to co-exit in the sadhya i.e. sound is a eternal because it has soundness. But the eternity exists in the Soul bereft of soundness. Therefore the reason "soundness" is not co-existent with the Sadhya. Further the co-existence of Sadhya with the hetu (reason) being the sub-character of invariable concomitance is not comprehended.

  (iii) Anupsanhārina (Inconclusive) - Here the reason that the Sadhya of which is not the counter-positive of the complete negation. Such fallacy debars the knowledge of absent-invariable concomitance.

(b) Virudha (Incompatible) – The hill has fire because of water. Here water is the counter positive of the negation pervading "fire" i.e. whenever there is fire, there is the negation of water.

(c) Āshraya-Asidha (Unproved property) – conclusion not warranted by the premises). It is either unproved property of the subject (Āshrayasiddha, or nature (swarupasiddha and invariable concomitance (Vyāpatvāsiddha).

(d) Satpratipaksha (Counter thesis) – sound is eternal because it has soundness – here it fails to argue that sound is not eternal because it is occasioned or generated. There is the fallacy of counter thesis.

(e) Bādhit (Contradictory) – "The pool contains fire" is rejected by the knowledge" the pond is fireless. So the negation of fire in the pool is not existing.

The Uses of Argumentation

(1) Argumentation and disputable management – when disputes spring up, we have several options. We may opt for not to take part in or prefer to run off. Literally an argument is taken for anti-

imperialism, anti-militarism, non-violence or non-aggression. Because we do use verbal language and not weapons to sort out our disputes. It is the advantageous motion to operate upon contentions or controversies.

(2) Cross-cultural viewpoint – It entails conceptual or thought supplement across cultures to observe factual scrutiny.

(3) Cognitive practice – It intends to enhance esoteric knowledge of gnoseology in several intrigues. But the exposition of the authentic conclusion maybe somewhat dubious in real life. We are but reluctant to deal conflicting belief or conception receptively. Such learned attitude to argument enables us to accomplish regulatory model for logical exercise.

(4) Concensus oriented – Since human beings are social animals who frequently collaborate with other ones to negotiate the problems effectively. So such motions are inspired by social problems and the assignment of social reciprocity to argumentative harmony.

(5) Anti-Pathetic symbiosis- genuine or real argumentative confrontation or collision may be anti-pathetic or pro-pathetic to various positions in many circumstances. Because a particular type of argument does not cope with all visions of arguments over different positions.

Thus argumentation (Vādaparamparā) is deduced as the trafficking of reasonings to validate averment on the crux of the submission in varied circumstances. It serves to be the most advantageous or flawless feedback in conflicting factors. It helps to cultivate self-supporting, expository and reproving reflections towards impartial conclusion.

It works as the juxtaposition or collation between the 02 groups of conviction demonstrating towards valid deduction

Argumentation and AL-I

Argumentation in AL-I was considered under the awareness or introspection of conception emerging from informal logic as argumentation progammes, setting circumstances, knobs of dialogue and logical motion. It can display head-on encounter-discussions in the court.

An argumentation in AL-I works on 03 major debates –

(1) The inner or essential tools of argument – where systematic motion investigates specifics antecedently, such as argument forte.

(2) Argumentation in multi-agent systems is to collocate sub-tract with its own dedicated conference series i.e. ArgMAx- being

operational that finds the argument to give the maximum value from a target function.

It is generally used to trace out the index-value in a set of elements corresponding to the maximum output score of a certain function.

(3) Application to specific task (Web-server) and domain – Where a set of connected Window computers sharing user account information or proposition to manage network administration in legal or valid argumentation.

Bar-Bench Intelligence

It should be borne in mind that Inference is focused on judgement. It is a tool for removal of confusion, indecision and dilemma through infallible debate on the logical subject (Pakshata) in question en route to valid injunctions.

The practice of discussion, debate and pleading on the given topic had been the special feature of oriental learning between the purva paksha (1st part or a group of party) and uttar paksha (Respondent party or a group of party) since time immemorial.

Today, a bit of its application is seen in GD (Group discussion). The same case has been practising in judicial system to conclude or determine judgement on disputed topics through logical arguments validated by claims, counter claims, sufficient reasons, evidences and acknowledgements.

Such inferential arguments are centred on assertions, reasonings evidence, common connotations, devices, mechanism, manoeuvre or rebuttal to critics.

In Courts the Bar and the Bench investigate the issues in depth (Deep Learning) by penta propositions or syllogism.

(1) **Pratigyā (Premises)**– According to Oriental "Nyaya-Siddhānta Muktavali's scientific verses – probandum is remarked with qualifier-qualified relations knowledgeable as the predicate. The antecedently determined part residing in the object is called-logical subject termed as probandum.

This 1st logical organ is essential to expose the claims for consideration between the parties. Only then the mediator (jury) is eager to investigate about the probans and adjudge integral and irresistible judgement.

(2) Hetu (Logical reason) – the proposition as reasoning catalyst treats the probans (smoke) and the probandum (fire) by the emergence of adjunctive qualifier.

Oriental philosopher Vāchaspati Mishra in his "Tātparyatika" explains as "Anuman hetuh" i.e. the reasoning of inferential evidence.

To be more clear, whatever we call **prior-determined invariable concomitant trait and conclusively characterized invariable concomitance** in the logical subject are inferential evidences in the council.

(3) Udāharan (Illustration) – the 3rd propositional illustration is needed to appease aspiration or eagerness or inquiry in a given situation. According to Vatsyāyana the propositional illustration indicates the characteristic Property of the both the probans and the probandum.

Nyayavartika-Critic-Vachaspati Mishra of Mithila has supported the view of Vatsyāyan indicating as the illustration being perceptual i.e. the perception of fire pervasion in the reasoning.

Nyayabhāsyakār Vatsyāyan explains that the characteristic property rests on the substance as in the case of external properties of colour, taste, smell, touch, measure and etc. Desiderative inquiry has characteristic property. And we prove the Soul in the fashion of will-based substance. So we are made to prove the utility of perceptual testimony on the above apperceptional entity (Self) in inferential system.

Since the Self (prān) is incorporeal (Nirākār) as the characteristic property within the substantive embodiment. So it is assumed by logical Inference or it is inferring.

Gangeshopādhyāya in "Tattva Chitamani" Avayava – has proved the legitimacy of inferential assertion after the perceptional definition on assuming the integrity of Inference subsisted by perception. Truly, when we inquire into inferential locus, we find that somehow or the other conviction of causality is effective through long lore.

(4) Upanaya (Analogy) – Gangeshopādhyāy treats Upanaya as the causative organ of the 3rd characteristic mark (Smoke, causing fire) – "Fire to be pervaded in the smoky hill in special case." Vashyakār Vatsyāyan treats Upanaya as analogous. The confirmation in special cause as there is smoke in the fire. Tatparyatikākār treats it also as indicator of analogical proposition. The knowledge by analogy is based on the recollection of perceptual proposition.

Nyaya Vrittikar (Critic) – Vishwanath panchānan - termed it to be the significance of cause-effect pervasion exclusively.

(5) Nigamana (Conclusive Judgement) – Here is the fire on the hill- Nyayasutra in **Anuman-chintāmani Avayava** defines it as repetition of the premises with reasoning instrument. Here all the 04-appellation is acknowledged as inter-consistents or reconciled or universal.

According to "Tarkasangrah Deepika" of Annam Bhatta – logical instruments are caused with intrinsic nature of the probandum (predicate).

According to Deepika Prakashakār's view – the recognized cause of proposition about the exclusive nature of the probandum stands for knowability of reasoning sense.

The very penta-organ of Inference has been elaborated in the preceding chapters exhaustively.

Here it is remarkable to note that the crux of issue should be adjudged on valid reasoning otherwise we might have erraneous, contradictory, reiterative or oxymoronic results–judgements. From the above version, I do infer that our Neo-logicians (Naiyāyika) innovated to optimise precise technical language in order to artifice Sanskrit-determinant phraseology for intended meaning. Since Nyāya language has been mechanized on logical interpretation on concept by great oriental logicians. Its determinant – determinated nuts-bolts in applied software technicalities do treat as Artificial language. That is why problems of Deep learning have been clarified and solved with creative techniques of determination of Navya-Nyaya. Navya-Nyaya-scientific methodology proves to be adroit or conversant in linguistics, legal thoughts, aestheties, statistics, satellite operation and other social needs.

The programming browser of Navya-Nyaya is to display technical terms in a pinpoint and clear cut manners to formulate logical languages for effective communication. Such innovative device helped to evolve the Neo-oriental school of thought. Further such contrivance helps to display universal consideration in clarity.

Since natural language being non-created may not be free from equivocation.

Further Navya-Nyaya's object-oriented language artificed with cognitive strategy or synergy works as an efficient device. Such coded language can be used as instrumental to steer IT sector through explicit application.

Such instrumental credit of Brain acuity or Brain HQ goes to Neo-logician – Gangeshopadhyaya displayed in "Tattvachintāmani".

An inferential system steers the matrix or network and encodes to fabricate legal arguments from the given propositions. Corroborative evidence serves as an instrument from hypotheses or speculations to the extent of judgement. Such step of argument prefers the logical phenomena. Now logical evidences or data demonstrations usually proceed with propositions standing for assumption. The objective is to intervene the premises on the code of Inference until the effect of conclusive judgement is suited.

Further causal Inference certifies logical conclusion which has been deducted or evidenced. The code of causal Inference is usually applied for evidence during a trial in lawsuit. Because causal Inference has logical conclusion or judgement delivered from valid exposition of evidences.

I have already discussed that Inference entailed cogent conclusion by observation and evaluation of the syllogistic proposition to be true or false.

Now the Intelligence (Information and evaluation), paves the way for evaluation with the process of judging the information to its context through sound or infallible sources. In other word Intelligence evaluation works on securing and reporting to information in order to manage to assist for reliable judgement.

Further all decisive judgements entail the application of our natural skill to evaluate or interpret data information which may be decomposed or degraded into a series of objections as follows –

Here the pros and cons analysis of the given problems as to the risk of success or failure depend on the spirit of analysis about. So our conscience regulates our analytical ability as follows:

(a) Analysis can apprise or deduce one as to how does moral or poor one's information/intelligence involve.

(b) It can deduce about the unknown or unnoticed matters or facts.

(c) It can deduce about where one needs to assess the position.

(d) It can deduce about where to investigate further.

(e) It can help to exchange briefing or know-how to the concerned.

In simple words practically –

(1) "Information" is a knowledge in basic data of any kind.

(2) 'Intelligence' is data efficiency to have been worked on accrual, worth or reliable sources.

(3) "Analysis" of either information or intelligence relates to

  (a) Classifying or degrading of substance into it constituent parts

  (b) Investigating determination of those parts.

  (c) Discerning the sources of substances to locate common canons behind them.

Data Intelligence and Evidence

Intelligence – can be applied to enforce law-operation. The exercise of intelligence convention concerning certain observation is normally an intervention in any evidence – panel.

The law would command- whether intelligence -instrument complied within the track of investigation is secured from any leakage of evidentiary information to let share automatically routine-evidentiary clues that would otherwise be available during scrutiny.

Evidence – Is the data of validity (Inference) to set up corroboration or testimony. Because NN (Navya Nyāya) is the data articulation to draw inference beyond the given information. Evidence is to investigate evaluation or interpretation which enables us to abstain from clone of trial and avert offending into impertinent or peripheral sectors. An evaluative skill should be applied at the initial stage of investigation if feasible in the consecutive steps as follows:-

(i) Compilation of planning – explicative approach to trace out the information required or sources of its availability.

(ii) Collation of data – corporation or combination of data into a layout and out of which they can be reclaimed or evaluated.

(iii) Analysis – judicious scrutiny of information to find out their interpretations and vital characters.

(iv) Evaluation – appraisal of credibility of the sources or characters or properties of the information.

(v) Promulgation or dispersion – delivering judgment on trial to the client.

Now I have come to know that intelligence and analysis play vital role to prescribe the source of enforcing rights or rectifying wrongs through pleading, practice, evidence; and representation of counsel, costs and allied matters as to where do the parties stand and show/why did the parties hit at the point.

The inferential analysis ripens the product of judgments with raw data and intelligence is a bid to tasking support to the client. Here we apply 02 types of intelligence –

(1) Cardinal or critical intelligence - It helps to analyze periodical changes as an abuse to the public protection. Further it evolves supervision of counter initiatives to foster programming enactment.

(2) Functional intelligence – It cares for investigating the team with hypotheses and inferences relating to certain parts of any unlawful or criminal operation. It undertakes to review illegal activities, their techniques, strength, limitations, vulnerability, and intentions to be effective for law enforcement moves.

Now inferential data - validity may be adjudged as follows–

(a) Unshakable – (i) by other self-sufficient means
(ii) deductive by itself
(iii) acceptable with other data on the locus.

(b) Probably true – ability to happen or exist and non-self-sufficient, but deductive by itself.

(c) Possibly true– Not self-sustaining and likely to be true but not certainly. Acceptable with other data to a limited extent.

(d) Problematically or ambiguously unshakable—
not self sustaining
not deductive
faithless at the time of encounter
despite its contigency.

(e) Debatable– proof available from the opposite party to be countered by the reasonable data on the locus.

Now the inferential mechanism provides viability to manoeuvre ML (Machine Learning) T.L. (Trial Learning) through valid reasonings or arguments. How does it work in intent is to be justified in the corpus of my technique-oriented approach to achieve the task from the bottom line of logical interpretation. We involve decisions being made about somethings in hand or what we observe. Here mind's phenomena consciousness and qualia – work with brain through neural chain to get at the conclusion or decision infallible. We know that our mind runs faster than light object as RAESR Automotive Tachyon (Hypothetical particle). The sensory motor-mind is susceptible to fluctuate in moments. So our valid concepts or notions on issues might be mistaken to suffer with 05 senses of input (hearing, touch, sight, taste and smell) in clumsy and chaotic conditions. So the motto of deductive logic is to amplify, investigate, assess and classify good and bad reasoning systems for Bio-Kingdom.

The celebrated logician Vachaspati Mishra of Mithila penned explanatory note on Nyaya Commentary of earlier logician Vātsyāyan. Wherein he expounded the science of legal reasoning or proceedings to conclude logical argument on the commotion of depute.

Further Alpha-Neo logician Gangeshopadhyaya of Mithila in his masterpiece "Tattvachintamani" elucidates the theory of logical reasoning as the instrumental cause to input penta-syllogism between the proponent (plaintiff) and the opponent (defendant) to assert their assumptive inference mediated for others.

Later on the umpire (Judge) on hearing pros and cons argument delivers sound or valid judgment tested to inferential mechanism on invariable concomitance and logical base by virtue of considerate character.

Further he systematically explores syllogism to infer the probandum, on Set – Algorithm under invariable concomitance and logical combo-reliance on associative algorithm – while explaining the quotient of syllogistic cognition to cause the distinct and distinction in plunging process. But his assumption was objected by oriental logician **Gadadhar Bhattacharya** to the extent of the subject (distinction) and the predicate (distinct) indicated by character-causing mind to conclude homogeneously. So he assumes a series of characteristic concomitance to be mental awareness and not verbal communication.

But celebrated logician Pakshadhar Mishra of Mithila counter-argues that there may be doubt of the validity of the proposition of the 1st party in the above situation. Such reasoning would debar the valid judgement. The conclusion of cognition between the probans and the probandum on synthetic character would hamper inferential judgement. Under the situation, the mediator (judge) has to adjudge with his own mental firmity over arguments or rejoinders posed by the parties.

Because our free and cool mind filters firm and fecund findings. However oriental logician Gadadhar Battacharya adds his own new opinion that verbal perception commensurates with compass of characteristic in re-applicable proposition.

In the circumstances, the verbal perception is reflected in the identification of essential character existing on cause-effect relationship between the logical subject and re-applicable proposition. The re-applicable proposition is the essential relation in reasoning part being the middle term. Reapplicable perception is not synthetic judgment rather conceivable in the mind.

Since where there is the essential relation apprehended to be assertained in the qualifier agent.

There is the proven apprehension of its qualified agent – viewed as reasoning (qualified) and invariable concomitance (qualifier), Thus the perception is that of it special type of determining adjectival concomitance. After conceiving this mind-driven cognition, the synthetic judgment itself is mind-like or cognitive.

Under weighing ordeal of balance (Tulādanda), there involves non-dubiety of authenticity or evidence in inferential conclusion or deductive knowledge viz. when we assess or compute weighing beam-parts for trading, we admit the particular measurement through inferential footing as justified assumption. We acknowledge or consider the universal knowledge with instrumental device without involvement of any doubt. Here doubt does not occur without any rhyme or reason. Because the emergence of doubt is trialed through inferential arguments.

Although law in logical intent is the admitted benchmark of right conduct. But judicial procedures are undertaken for conceptual pabulum.

With the momentum of times the network access on figural facts or dossiers with intelligence data is needed for legal conclusion.

The solicitation of AL-I (Artificial Intelligence) in legal judgment may furnish a reference to solve impugned disputes with digital or programmed genre.

Judicial systems are soliciting AL-9 to evaluate and analyze vast amount of legal data to streamline the case-laws or case-history. Such mechanism of man-machine fusion helps the judges to predict on issues in order to speed up case-trial without immoderate delay.

Due to physiological constraints of human acumen in judgemental faculty, we may not monitor the whole ugly factors overhanging our social solidarity.

So to control and condemn the rapidly ever-immersing mood of the land, the judicial authority should apply Robotic device for our timely reliefs.

We are the locus of Anumān (inference). Inference and arguments are 02 sides of the same coin. Arguments tend to be Anuman (Inference) and Anuman tends to be an argument.

Further argument is the character of Anuman processed through rational inquiries. Whereas Anumān depends on specific logical concepts called logical qualifiers.

In case-trial, knowledge-based reasoning is exercised for logical conclusion predicted from available antecedent knowledge (Smoke in the hill). When we compile or acquire information to understand them. And the information being understood we learn to apply our knowledge for judgment which is based on –

(1) Gyān satyāpan or Yathārtha Gyān (knowledge validation) – a critico-legal study on rational analysis to review evidence and device in the knowledge-base (Pakshata).

Such procedure is done at 02 levels –

(a) Universal – this model comes from interlinks among the key-factors of the KV (Knowledge Validation) norms

(b) Certain principles – for KR (Knowledge Representation) rules, figures and neural networks.

In Court cabin or room the Bar and the Bench inductively infer new decisive opinion from antecedent knowledge. Where arguments are based on inductive logic to conclude reasonable, probable, credible and feasible pace towards valid observation. Systems in AL-I depend on Mega-volume of training data to adeptly monitor tasks. When DL (Deep Learning) devices have been capable to embody brain-causing inference-mechanism in AL-I programme.

Human laws are prescriptive which would/should be descriptive to analyze evaluate, interpret and explore for truth-abiding decision called-synthetic judgment.

Such cause entails merger of human intelligence and AL-I turns computer into humanoid. Further the blended softwares equip computing models to integrate visual or natural language processing-just as human brain does for comprehensive communication.

Further AL-I tends to be mechanised with the following approaches –

(1) AL-I aims to capture the knowledge of human experts to support a decision-making. Some systems encode expert knowledge as rules and are, therefore, referred to as rule-based systems.

(2) Case-based approach of reasoning, substitutes and cases for rules.

(3) AL-I tends to delineate KR (Knowledge Representation). Basically it is the study of how beliefs of objectival judgment of an intelligence-agent can be demonstrated suitably for automated reasoning.

(4) AL-I subsumes finding from logic to automated multi-reasoning Mechanism such as –

Application of rules or relation of sets or subsets.

In KR, Inference-languages have Inference-Engine (IE) a a part of the system. Here the KR caches- processed through arguments as follows –

(a) Fundamental conception of intelligence-reasoning.

(b) Ontological commitment.

(c) Fragmentary theory of intelligence-reasoning.

(d) Set of inferences and KR urges under valid options.

The rapidly developing social needs warrant optimum auto-solution beyond biological limitations.

To expedite Generative AL-I being innovative potentials to lodge Robotic AL-I economy in social arena, we would predict about its ethical scenario. Although many machine-predictions are being deployed to assist in decision-making get rectified by human explanation eventually.

Programming ethics into Machine may be intricate. So we must proceed carefully lest incur invalid judgment. Mind that AL-I agents are still far away from being able to replace human judgments in complexity.

A venture to codify ethics into rules whether for Robots or Bio-tools or complex driven-judgment-reveals the basic challenges in rendering human-moral complexity into Algorithmic device.

Here programming ethics into Machine Learning (ML) and AL-I are more complicated. Although Artificial Moral Agents may work in –

(i) Ethical machine consequence like-Robot Jockeys.

(ii) Implicit ethical agents – Machine with safety guidelines – as auto-pilot systems.

(iii) Explicit ethical agents – going beyond set rules with social responsibility.

(iv) Full ethical agents – capable of justifying ethical explanations.

Update appreciation argues that AL-I system, Developer of AL-I or official dependence on AL-I data may not be liable to incur immoral or unethical fall to the extent of punishing against wrong doings.

Normally, all humanoid models and human beings do incur mistakes in decision-making. Whereas human beings are afraid of being punished mentally or physically. So they remain alert against law-breakings. At the same time the machine-made decisions are not afraid of being punished emotionally or effectively.

In oriental domain, legal systems bring about Rules of Law through "Intuitive Telepathy" (6th sense being next to 05 ones) against "hypothesis" or "assumption". So we, orientalists, should prescribe the moral-based law enactment to achieve material and immaterial integrity. But we are sailing in the twin-stream of alien and own legal systems being The Providence of "Paradise Lost".

हम लोग काष्ठ प्रयुक्त पाकशाला में अनादि काल से प्राच्यानुमान आधारित स्थाली पुलाकन्याय के अंतर्गत यादृच्छिक प्रतिचयन एकल प्रणाली से अनुभूत प्रयोग के द्वारा कुल तण्डुल सिद्धि का पश्च ज्ञान प्राप्त करते आ रहे हैं। अभी प्राच्यानुमानाधारित मानव मस्तिष्क की अंतर्बोध चेतना के अनुकरणार्थ स्वचालित - कुशल तंत्रिका यंत्र के द्वारा जटिल कार्य का निष्पादन हो रहा है। साथ ही मानव समाज में इस विशिष्ठ तंत्रिका यंत्र के द्वारा प्राच्यन्याय-प्रदान-प्रणाली से अव्यविचार का भी सुलभ संचरण होगा। डॉ मिश्र की इस शोधप्रद प्रस्तुति में प्रायोगिक वैज्ञानिकता है।

By Patron of the Programming